Carpentry and Joinery

NVQ and Technical Certificate **Level 3**

Heinemann is an imprint of Harcourt Education Limited, a company incorporated in England and Wales, having its registered office: Halley Court, Jordan Hill, Oxford OX2 8EJ. Registered company number: 3099304

www.harcourt.co.uk

Heinemann is the registered trademark of Harcourt Education Limited

Text © Carillion Construction Ltd

First published 2007

12 11 10 09 08 07
10 9 8 7 6 5 4 3 2 1

British Library Cataloguing in Publication Data is available from the British Library on request.

ISBN 978 0 435464 71 4

Copyright notice

Edited by Sarah Christopher
Designed by HL Studios
Typeset by HL Studios
Printed in the UK by Scotprint

Illustrated by HL studios
Original illustrations © Harcourt Education Limited 2007

Cover design by GD Associates
Cover photo Harcourt Ltd © Gareth Boden

Websites

The websites used in this book were correct and up to date at the time of publication. It is essential for tutors to preview each website before using it in class so as to ensure that the URL is still accurate, relevant and appropriate. We suggest that tutors bookmark useful websites and consider enabling students to access them through the school/college intranet.

Contents

Acknowledgements

Photos

Harcourt would like to thank the following for their kind permission to reproduce photographs in this book.

Alamy Images/Adrian Sherratt, p 237; Alamy Images/Andrew Paterson, p 81; Alamy Images/Imagebroker, p 223; Alamy Images/Justin Kase, p 7; Alamy Image/Nic Hamilton, p 3; Alamy Images/David R. Frazier Photolibrary Inc, p 85 (Taped); Alamy Images/The Photolibrary Wales, p14; Alamy Images/Trish Grant, p 187; Construction Photography/Chris Henderson, p 9; Construction Photography/CP Stock, p 43; Construction Photography/Darren Holden, p 1; Construction Photography/DIY Photolibrary, p 40, 84; Construction Photography/Grant Smith, p 12; Construction Photography/Paul McMullin, p 39, 41 (Plastering), p 169; Construction Photography/Xavier de Canto, p 17; Corbis, p 55, p 117; Corbis/Creasource, p 27; Corbis/Helen King, p 97; Corbis/Holger Schreibe/Zefa, p 104; Corbis/Martin Meyer/Zefa, p 42; Getty Images, p 209; Harcourt Ltd/Gareth Boden, p 81 (Steps 1-4), p 82 (all), p105 (hinge), p124 (all), p 156 (all), p 159-60 (all), 210, p 211 (both), 218, p 219; Harcourt Ltd/Jules Selmes, p 69, 81 (Step 5), p 86-7 (all), p 88 (all), 102-3 (all), 105 (mortise lock/plates), p 106 (all), p 170, p 171 (all), p 172, p 183, p 190 (flatting), p 194; Photographers Direct/Bjorn Beheydt, p 25; Photographers Direct/David Griffiths, p 23; Science Photo Library/Garry Watson, p 8 (corrosive); Startrite, p 191, p 199, p 202, p 206; Toolbank, p 218; Topham Picturepoint, p 8 (toxic & explosion)

Introduction

This book has been written based on a concept used in Carillion Training Centres for many years. That concept is about providing learners with the necessary information they need to support their studies and at the same time ensuring it is presented in a style, which they find both manageable and relevant.

The contents of this book have been put together by Kevin Jarvis who has a wealth of knowledge and experience in both training for NVQ and technical certificates in his trade.

This book builds upon material covered in *Carpentry and Joinery NVQ and Technical Certificate Level 2*, which introduced readers to the principles behind many of the carpentry topics that will be explored in greater depth in this book.

This book has been produced to help the learner build a sound knowledge and understanding of all aspects of the NVQ and Technical Certificate requirements associated with their trade. It has also been designed to provide assistance when revising for Technical certificate end tests and NVQ job knowledge tests.

Each chapter of this book relates closely to a particular unit of the NVQ or Technical Certificate and aims to provide just the right level of information needed to form the required knowledge and understanding of that subject area.

This book provides a brief introduction to the supervisory role that a Level 3 student will be working towards as well as providing the knowledge on the tools, materials and methods of work allowing you to complete work activities safely, effectively and productively. Upon completion of your studies, this book will remain a valuable source of information and support when carrying out your work activities.

For further information on how the contents of this book matches to the unit requirements of the NVQ and Advanced Construction Award, please use the detailed mapping document which can be found on our website www.heinemann.co.uk

How this book can help you

You will discover a variety of features throughout this book, each of which have been designed and written to increase and improve your knowledge and understanding. These features are:

- **Photographs** – many photographs that appear in this book are specially taken and will help you to follow a step-by-step procedure or identify a tool or material.

- **Illustrations** – clear and colourful drawings will give you more information about a concept or procedure.

- **Definitions** – new or difficult words are picked out in bold in the text and defined in the margin.

- **Remember** – key concepts or facts are highlighted in these margin boxes.

- **Find out** – carry out these short activities and gain further information and understanding of a topic area.

- **Did you know?** – interesting facts about the building trade.

- **Safety tips** – follow the guidance in these margin boxes to help you work safely.

- **FAQs** – frequently asked questions appear in all chapters along with informative answers from the experts.

- **On the job scenarios** – read about a real-life situation and answer the questions at the end. What would you do? (Answers can be found on the Tutor Resource Disk that accompanies this book.)

- **End of chapter Knowledge checks** – test your understanding and recall of a topic by completing these questions.

- **Glossary** – at the end of this book you will find a comprehensive glossary that defines all the **bold** words and phrases found in the text. A great quick reference tool.

chapter 1

Health and safety

OVERVIEW

The construction industry is one of the most dangerous industries in the UK. Every year over 100 people are killed and thousands seriously injured, while thousands more suffer health problems such as **dermatitis**, **asbestosis** and deafness. You can see why learning as much as you can about health and safety is very important.

Level 2 gave a good grounding in health and safety and informed you of what you need to know and do. This book takes things a step further: as a Level 3 candidate working in a more supervisory role, your responsibilities and actions will be different.

This chapter will cover:

- health and safety legislation
- health and welfare
- risk assessments.

Health and safety legislation

Definition

Hazardous – dangerous or unsafe

While at work, whatever your location or type of work, you need to be aware that there is important **legislation** you must comply with. Health and safety legislation is there not just to protect you, but also states what you must and must not do to ensure that no workers are placed in a situation hazardous to themselves or others.

Each piece of legislation covers your own responsibilities as an employee and those of your employer – it is vital that you are aware of both. As a Level 3 candidate, you not only have to think of your responsibilities for you own actions, but must also consider your supervisory responsibilities for others. These may involve ensuring that others are aware of legislation and considering such legislation when you are overseeing others' work.

What is legislation?

Legislation means a law or set of laws passed by Parliament, often called an Act. There are hundreds of Acts covering all manner of work from hairdressing to construction. Each Act states the duties of the **employer** and **employee**. If an employer or employee does something they shouldn't – or doesn't do something they should – they can end up in court and be fined or even imprisoned.

Approved code of practice, guidance notes and safety policies

Definition

Proactive – acting in advance, before something happens (such as an accident)

Reactive – acting after something happens, in response to it

As well as Acts, there are two sorts of codes of practice and guidance notes: those produced by the **Health and Safety Executive (HSE)**, and those created by companies themselves. Most large construction companies – and many smaller ones – have their own guidance notes, which go further than health and safety law. For example, the law states that that everyone must wear safety boots in a hazardous area, but a company's code may state that everyone must wear safety boots at all times. This is called taking a **proactive** approach, rather than a **reactive** one.

Most companies have some form of **safety policy** outlining the company's commitment and stating what they plan to do to ensure that all work is carried out as safely as possible. As an employee, you should make sure you understand the company's safety policy as well as their codes of practice. If you act against company policy you may not be prosecuted in court, but you could still be disciplined by the company or even fired.

When you are acting as a supervisor you will need to ensure that staff you are supervising understand the safety policy too. The safety policy may even require you to take on further responsibilities as a supervisor, such as running safety drills and checks: if so you will need to understand what is involved in these and ask for any necessary support or training from your employer.

Health and safety legislation you need to be aware of

There are some 20 pieces of legislation you will need to be aware of, each of which sets out requirements for employers and often employees. One phrase often comes up here – '*so far as is reasonably practicable*'. This means that health and safety must be adhered to at all times, but must take a common sense, practical approach.

For example, the Health and Safety at Work Act 1974 states that an employer must *so far as is reasonably practicable* ensure that a safe place of work is provided. Yet employers are not expected to do everything they can to protect their staff from lightning strikes, as there is only a 1 in 800,000 chance of this occurring – this would not be reasonable!

We will now look at the regulations that will affect you most.

The Health and Safety at Work Act (HASAW) 1974

HASAW applies to all types and places of work and to employers, employees, the self-employed, sub-contractors and even suppliers. The act is there to protect not only the people at work but also the general public, who may be affected in some way by the work that has been or will be carried out.

Legislation is there to protect employees and the public alike

The main **objectives** of the health and safety at work act are to:

* ensure the health, safety and welfare of all persons at work

* protect the general public from all work activities

* control the use, handling, storage and transportation of explosives and highly flammable substances

* control the release of noxious or offensive substances into the atmosphere.

To ensure that these objectives are met there are duties for all employers, employees and suppliers.

Employer's duties

Employers must:

- provide safe **access** and **egress** to and within the work area
- provide a safe place to work
- provide and maintain plant and machinery that is safe and without risks to health
- provide information, instruction, training and supervision to ensure the health and safety at work of all employees
- ensure safety and the absence of risks to health in connection with the handling, storage and transportation of articles and substances
- have a written safety policy that must be revised and updated regularly, and ensure all employees are aware of it
- involve trade union safety representatives, where appointed, in all matters relating to health and safety
- provide and not charge for **personal protective equipment (PPE)**.

Employee's duties

The employee must:

- take reasonable care for his/her own health and safety
- take reasonable care for the health and safety of anyone who may be affected by his/her acts or **omissions**
- co-operate with his/her employer or any other person to ensure legal **obligations** are met
- not misuse or interfere with anything provided for their health and safety
- use any equipment and safeguards provided by his/her employer.

Those supervising others will need to make sure that staff are using all safety equipment in the appropriate way, and are taking any necessary steps to ensure that they do not jeopardise their own or others' safety.

Employees cannot be charged for anything that has been done or provided for them to ensure that legal requirements on health and safety are met. The self-employed and sub-contractors have the same duties as employees – and if they have employees of their own, they must obey the duties set down for employers.

Definition

Access – entrance, a way in

Egress – exit, a way out

PPE – personal protective equipment, such as gloves, a safety harness or goggles

Definition

Omission – something that has not been done or has been missed out

Obligation – something you have a duty or a responsibility to do

Supplier's duties

Persons designing, manufacturing, importing or supplying articles or substances for use at work must ensure that:

- articles are designed and constructed so that they will be safe and without risk to health at all times while they are being used or constructed

- substances will be safe and without risk to health at all times when being used, handled, transported and stored

- tests on articles and substances are carried out as necessary

- adequate information is provided about the use, handling, transporting and storing of articles or substances.

HASAW, like most of the other acts mentioned, is enforced by the Health and Safety Executive (HSE). HSE inspectors visit sites and have the power to:

- enter any premises at any reasonable time

- take a police constable with them

- examine and investigate anything on the premises

- take samples

- take possession of any dangerous article or substance

- issue improvement notices giving a company a certain amount of time to sort out a health and safety problem

- issue a **prohibition** notice stopping all work until the site is deemed safe

- **prosecute** people who break the law including employers, employees, self-employed, manufacturers and suppliers.

Provision and Use of Work Equipment Regulations 1998 (PUWER)

These regulations cover all new or existing work equipment – leased, hired or second-hand. They apply in most working environments where the HSW applies, including all industrial, offshore and service operations.

PUWER covers starting, stopping, regular use, transport, repair, modification, servicing and cleaning.

'Work equipment' includes any machinery, appliance, apparatus or tool, and any assembly of components that are used in non-domestic premises. Dumper trucks, circular saws, ladders, overhead projectors and chisels would all be included, but substances, private cars and structural items all fall outside this definition.

The general duties of the act require equipment to be:

- suitable for its intended purpose and only to be used in suitable conditions
- maintained in an efficient state and maintenance records kept
- used, repaired and maintained only by a suitably trained person, when that equipment poses a particular risk
- able to be isolated from all its sources of energy
- constructed or adapted to ensure that maintenance can be carried out without risks to health and safety
- fitted with warnings or warning devices as appropriate.

In addition, the act requires:

- all those who use, supervise or manage work equipment to be suitably trained
- access to any dangerous parts of the machinery to be prevented or controlled
- injury to be prevented from any work equipment that may have a very high or low temperature
- suitable controls to be provided for starting and stopping the work equipment
- suitable emergency stopping systems and braking systems to be fitted to ensure the work equipment is brought to a safe condition as soon as reasonably practicable
- suitable and sufficient lighting to be provided for operating the work equipment.

Control of Substances Hazardous to Health Regulations 2002 (COSHH)

These regulations state how employees and employers should work with, handle, store, transport and dispose of potentially hazardous substances (substances that might negatively affect your health) including:

- substances used directly in work activities (e.g. adhesives or paints)
- substances generated during work activities (e.g. dust from sanding wood)
- naturally occurring substances (e.g. sand dust)
- biological agents (e.g. bacteria).

Hazardous substances

These substances can be found in nearly all work environments. All are covered by COSHH regulations except asbestos and lead paint, which have their own regulations.

To comply with COSHH regulations, eight steps must be followed:

Step 1 Assess the risks to health from hazardous substances used or created by your activities.

Step 2 Decide what precautions are needed.

Step 3 Prevent employees from being exposed to any hazardous substances. If prevention is impossible, the risk must be adequately controlled.

Step 4 Ensure control methods are used and maintained properly.

Step 5 Monitor the exposure of employees to hazardous substances.

Step 6 Carry out health **surveillance** to ascertain if any health problems are occurring.

Step 7 Prepare plans and procedures to deal with accidents such as spillages.

Step 8 Ensure all employees are properly informed, trained and supervised.

Safety tip

Not all substances are labelled, and sometimes the label may not match the contents. If you are in any doubt, do not use or touch the substance

Identifying a substance that may fall under the COSHH regulations is not always easy, but you can ask the supplier or manufacturer for a COSHH data sheet, outlining the risks involved with it. Most substance containers carry a warning sign stating whether the contents are corrosive, harmful, toxic or bad for the environment.

As you can see, supervision is specifically mentioned in Step 8 above. For the Level 3 candidate, this may mean a much greater and more active involvement in compliance with COSHH regulations: you may become a doer – carrying out training, overseeing others and even carrying out some of the previous steps on behalf of your employer – rather than just a receiver of information and guidance.

Common safety signs for corrosive, toxic and harmful materials

The Personal Protective Equipment at Work Regulations 1992 (PPER)

These regulations cover all types of PPE, from gloves to breathing apparatus. After doing a risk assessment and once the potential hazards are known, suitable types of PPE can be selected. PPE should be checked prior to issue by a trained and competent person and in line with the manufacturer's instructions. Where required, the employer must provide PPE free of charge along with a suitable and secure place to store it.

The employer must ensure that the employee knows:

- the risks the PPE will avoid or reduce
- its purpose and use
- how to maintain and look after it
- its limitations.

The employee must:

- ensure that they are trained in the use of the PPE prior to use
- use it in line with the employer's instructions
- return it to storage after use
- take care of it, and report any loss or defect to their employer.

Remember

PPE must only be used as a last line of defence

As with many areas of health and safety, as a Level 3 candidate you may find yourself in a position somewhere between employer and employee, supervising the work of others where PPE is in use. As a supervisor, you may not carry any direct legal responsibilities under the PPER legislation, but your employer would still be looking for you to support them in meeting these obligations, and would be looking to you to set the standards for others. At the same time, you would be obliged to look after yourself as an employee under the terms of the legislation and keep yourself safe!

The Control of Noise at Work Regulations 2005

At some point in your career in construction, you are likely to work in a noisy working environment. These regulations help protect you against the consequences of being exposed to high levels of noise, which can lead to permanent hearing damage.

Damage to hearing has a range of causes, from ear infections to loud noises, but the regulations deal mainly with the latter. Hearing loss can result from one very loud noise lasting only a few seconds, or from relatively loud noise lasting for hours, such as a drill.

The regulations state that the employer must:

- assess the risks to the employee from noise at work

- take action to reduce the noise exposure that produces these risks

- provide employees with hearing protection or, if this is impossible, reduce the risk by other methods

- make sure the legal limits on noise exposure are not exceeded

- provide employees with information, instruction and training

- carry out health surveillance where there is a risk to health.

Anyone who is supervising the work of others may be asked to (or may wish to) monitor noise levels in the work area and advise the employer if any problems arise from excessive noise exposure.

Noise at work

Did you know?

Noise is measured in **decibels (dB)**. The average person may notice a rise of 3dB, but with every 3dB rise, the noise is doubled. What may seem like a small rise is actually very significant

The Work at Height Regulations 2005

Construction workers often work high off the ground, on scaffolding, ladders or roofs. These regulations make sure that employers do all that they can to reduce the risk of injury or death from working at height.

The employer has a duty to:

- avoid work at height where possible

- use any equipment or safeguards that will prevent falls

- use equipment and any other methods that will minimise the distance and consequences of a fall.

As an employee, you must follow any training given to you, report any hazards to your supervisor and use any safety equipment made available to you. As a supervisor, you would need to understand the duties of both employer and employee, being aware that staff under your supervision should never be put in danger by failing to meet the standards required.

The Electricity at Work Regulations 1989

These regulations cover any work involving the use of electricity or electrical equipment. An employer has the duty to ensure that the electrical systems their employees come into contact with are safe and regularly maintained. They must also have done everything the law states to reduce the risk of their employees coming into contact with live electrical currents.

The Manual Handling Operations Regulations 1992

These regulations cover all work activities in which a person does the lifting rather than a machine. They state that, wherever possible, manual handling should be avoided, but where this is unavoidable, a risk assessment should be done.

In a risk assessment, there are four considerations:

- *Load* – is it heavy, sharp-edged, difficult to hold?

- *Individual* – is the individual small, pregnant, in need of training?

- *Task* – does the task require holding goods away from the body, or repetitive twisting?

- *Environment* – is the floor uneven, are there stairs, is it raining?

After the assessment, the situation must be monitored constantly and updated or changed if necessary.

The Reporting of Injuries, Diseases and Dangerous Occurrences Regulations 1995 (RIDDOR)

Under RIDDOR, employers have a duty to report accidents, diseases or dangerous occurrences. The HSE use this information to identify where and how risk arises and to investigate serious accidents.

Other acts to be aware of

You should also be aware of the following pieces of legislation:

- The Fire Precautions (Workplace) Regulations 1997
- The Fire Precautions Act 1991
- The Highly Flammable Liquids and Liquid Petroleum Gases Regulations 1972
- The Lifting Operations and Lifting Equipment Regulations 1998
- The Construction (Health, Safety and Welfare) Regulations 1996
- The Environmental Protection Act 1990
- The Confined Spaces Regulations 1997
- The Working Time Regulations 1998
- The Health and Safety (First Aid) Regulations 1981
- The Construction (Design and Management) Regulations 1994.

You can find out more at the library or online.

Find out

Look into the other regulations listed here via the Government website www.hse.gov.uk

Health and welfare

As a worker in the construction industry, you will be at constant risk unless you adopt a good health and safety attitude. By following the rules and regulations, and by taking reasonable care of yourself and others – especially where acting as a supervisor – you will become a safe worker and reduce the chance of injuries and accidents. Given the statistics on safety, the supervisor's role is crucial here: few other people will be a better position to understand the day-to-day work of a site, be in touch with those doing the labour and spot 'danger points' where accidents or ill health could occur.

The two most common risks to a construction worker

Accidents

We often hear the saying 'accidents will happen', but in the construction industry, the truth is that most accidents are caused by human error – someone does something they shouldn't or, just as importantly, does not do something they should. Accidents often happen when someone is hurrying, not paying attention, trying to cut corners or costs, or has not received the correct training.

If an accident happens, you or the person it happened to may be lucky enough to escape uninjured. More often, an accident will result in an injury, whether minor (e.g. a cut or a bruise), major (e.g. loss of a limb) or even fatal. The most common causes of fatal accidents in the construction industry are:

- falling from scaffolding
- being hit by falling materials
- falling through fragile roofs
- being hit by a forklift or lorry
- **electrocution**.

Ill health

In the construction industry, you will be exposed to substances or situations that may be harmful to your health. Some of these health risks may not be noticeable straight away and it may take years for symptoms to be noticed and recognised. Ill health can result from:

- exposure to dust (e.g. asbestos) – breathing problems and cancer
- exposure to **solvents** or chemicals – dermatitis and other skin problems
- lifting heavy or difficult loads – back injury and pulled muscles
- exposure to loud noise – hearing problems and deafness
- using vibrating tools – **vibration white finger** and other hand problems.

A fall could be fatal

Remember

Everyone has a responsibility for health and safety, but accidents and health problems still happen too often. Make sure you do what you can to prevent them

Staying healthy

As well as watching for hazards, you must also look after yourself and stay healthy.

One of the easiest ways to do this is to wash your hands regularly: this prevents hazardous substances entering your body through ingestion (swallowing). You should always wash your hands after going to the toilet and before eating or drinking.

You can also make sure that you wear barrier cream and the correct PPE, and only drink water that is labelled as drinking water.

Welfare facilities

Welfare facilities are things that an employer must provide to ensure a safe and healthy workplace.

- **Toilets** – The number provided depends on how many people are intending to use them. Males and females can use the same toilet providing the door can be locked from the inside. Toilets should ideally be flushable with water or, if this is not possible, with chemicals.

- **Washing facilities** – Employers must provide a basin large enough for people to wash their hands, face and forearms, with hot and cold running water, soap and a way to dry your hands. Showers may be needed if the work is very dirty or if workers are exposed to toxic or corrosive substances.

- **Drinking water** – A supply of clean drinking water should be available, from a mains-linked tap or bottled water. Mains-linked taps need to be clearly labelled as drinking water; bottled drinking water must be stored where there is no chance of contamination.

- **Storage or dry room** – Every building site must have an area where workers can store clothes not worn on site, such as coats and motorcycle helmets. If this area is to be used as a drying room, adequate heating must be provided.

- **Lunch area** – Every site must have facilities for taking breaks and lunch well away from the work area. There must be shelter from the wind and rain, with heating as required, along with tables and chairs, a kettle or urn and a means of heating food.

Risk assessments

You will have noticed that most of the legislation we have looked at requires risk assessments to be carried out. The Management of Health and Safety at Work Regulations 1999 require every employer to make suitable and sufficient assessment of:

- the risks to the health and safety of his/her employees to which they are exposed while at work

- the risks to the health and safety of persons not in his/her employment arising out of or in connection with his/her work activities.

Remember

Some health problems do not show symptoms straight away – and what you do now can affect you greatly in later life

Did you know?

We all carry out risk assessments hundreds of times a day. For example, every time we boil a kettle, we do a risk assessment without even thinking about it: for example, by checking the kettle isn't too full, or the cable frayed, and by keeping children out of the way

Definition

Making a risk assessment – measuring the dangers of an activity against the likelihood of accidents taking place.

As a Level 3 candidate, it is vital that you know how to carry out a risk assessment. Often you may be in a position where you are given direct responsibility for this, and the care and attention you take over it may have a direct impact on the safety of others. You must be aware of the dangers or hazards of any task, and know what can be done to prevent or reduce the risk.

There are five steps in a risk assessment – here we use cutting the grass as an example:

Step 1 Identify the hazards

When cutting the grass the main hazards are from the blades or cutting the wire, electrocution and any stones that may be thrown up.

Even an everyday task like cutting the grass has its own dangers

Step 2 Identify who will be at risk

The main person at risk is the user but passers-by may be struck by flying debris.

Step 3 Calculate the risk from the hazard against the likelihood of an accident taking place

The risks from the hazard are quite high: the blade or wire can remove a finger, electrocution can kill and the flying debris can blind or even kill. The likelihood of an accident happening is medium: you are unlikely to cut yourself on the blades, but the chance of cutting through the cable is medium, and the chance of hitting a stone high.

Step 4 Introduce measures to reduce the risk

Training can reduce the risks of cutting yourself; training and the use of an **RCD** can reduce the risk of electrocution; and raking the lawn first can reduce the risk of sending up stones.

Step 5 Monitor the risk

Constantly changing factors mean any risk assessment may have to be modified or even changed completely. In our example, one such factor could be rain.

Definition

RCD – residual current device, a device that will shut the power down on a piece of electrical equipment if it detects a change in the current, thus preventing electrocution

On the job: Causing a hazard

Danielle and Stephanie are working on a building site laying upper floor chipboard flooring, with Danielle acting as supervisor. At lunchtime they both rush off leaving the area unsupervised. Roger walks over to borrow a nail gun and when he steps on what seems to be a fixed board, the board flips up. Roger falls down to the next level, seriously injuring himself. Who do you think is to blame? What would you have done in the circumstances?

FAQ

What is the difference between being an employee and being a supervisor when it comes to health and safety?

As a supervisor, you need to be aware of the different pieces of legislation and consider them when you are overseeing other people's work. You may be given responsibility to help your employer comply with their part of legislation too.

How do I find out what safety legislation is relevant to my job?

Ask your employer or contact the HSE at www.hse.gov.uk.

How do I find my company's safety policy?

Ask your supervisor or employer.

When do I need to do a risk assessment?

A risk assessment should be carried out if there is any chance of an accident happening. To be on the safe side, you should make a risk assessment before starting each task.

Do I need to read and understand every regulation?

No. It is part of your employer's duty to ensure that you are aware of what you need to know.

Knowledge check

1. What is legislation?

2. What is an approved code of practice?

3. What is the purpose of a safety policy?

4. What does 'so far as is reasonably practicable' mean?

5. Who enforces the health and safety regulations?

6. State three of the main objectives of the Health and Safety at Work Act 1974.

7. State four duties included in the PUWER regulations.

8. List three things that can cause ill health, and what health problems they can create.

9. State five welfare facilities that *must* be made available.

10. What is the purpose of a risk assessment?

11. Name three ways in which a supervisor's role with health and safety differs from an ordinary employee?

chapter 2

Building documentation

OVERVIEW

In the construction industry you come across a wide range of documentation, and as a Level 3 apprentice you will encounter different types of documents more frequently. This chapter covers the main building documentation you will see, explaining what each type of documentation is and what it is used for. The types of documentation covered in this chapter are:

- plans and drawings
- contract documents
- Building Regulations documentation
- general site paperwork.

Plans and drawings

Plans and drawings are vital to any building work as a way of expressing the client's wishes. Drawings are the best way of communicating a lot of detailed information without the need for pages and pages of text. Drawings form part of the contract documents (which will be explained later) and go through several stages before they are given to tradespeople for use.

Stage 1 The client sits down with an architect and explains his/her requirements.

Stage 2 The architect produces drawings of the work and checks with the client to see if the drawings match what the client wants.

Stage 3 If required, the drawings go to planning to see if they can be allowed, and are also scrutinised by **the Building Regulations** Authority. It is at this stage that the drawings may need to be altered to meet Planning or Building Regulations.

Stage 4 Once passed, the drawings are given to contractors along with the other contract documents, so that they can prepare their tenders for the contract.

Stage 5 The winning contractor uses the drawings to carry out the job. At this point the drawings will be given to you to work from.

There are three main types of working drawings: location drawings, component drawings and assembly drawings. We will look at each of these in turn.

Location drawings

Location drawings include:

- **block plans**, which identify the proposed site in relation to the surrounding area. These are usually drawn at a scale of 1:2500 or 1:1250

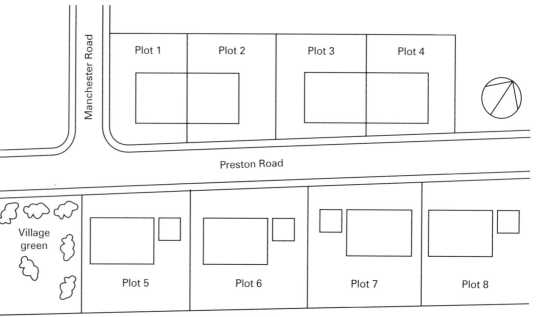

Figure 2.1 Block plan

Definition

The Building Regulations – a set of regulations brought in to deal with poor housing conditions, which now restrict what can be built, how and where. For more details, see Chapter 3, pages 40–42

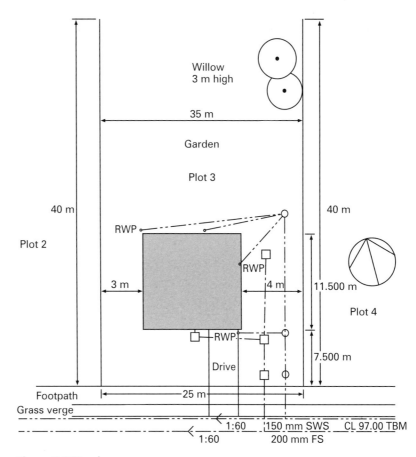

Figure 2.2 Site plan

- **site plans**, which give the position of the proposed building and the general layout of things such as services and drainage. These are usually drawn at a scale of 1:500 or 1:200

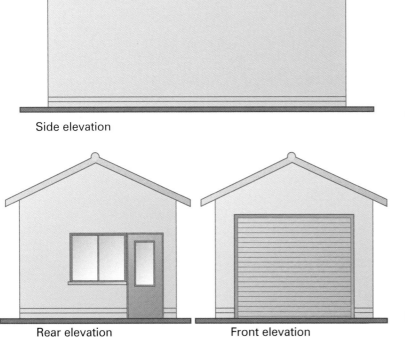

- **general location drawings**, which show different elevations and sections of the building. These are usually drawn at a scale of 1:200, 1:100 or 1:50

Figure 2.3 General location drawing

Component drawings

Component drawings include:

- **range drawings**, which show the different sizes and shapes of a particular range of components. These are usually drawn at a scale of 1:50 or 1:20

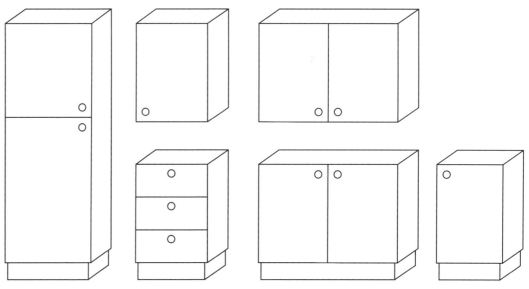

Figure 2.4 Range drawing

- **detailed drawings**, which show all the information needed to complete or manufacture a component. These are usually drawn at a scale of 1:10, 1:5 or 1:1.

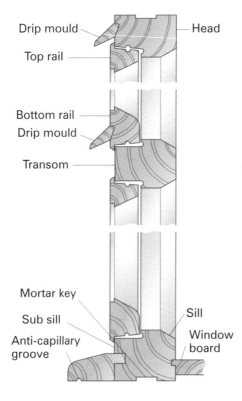

Drip mould — — Head
Top rail —
Bottom rail —
Drip mould —
Transom —
Mortar key —
Sub sill —
Sill
Window board
Anti-capillary groove

Figure 2.5 Detailed drawing

Assembly drawings

Assembly drawings are similar to detailed drawings and show in great detail the various joints and junctions in and between the various parts and components of a building. Assembly drawings are usually drawn at a scale of 1:20, 1:10 or 1:5.

All plans and drawings contain symbols and abbreviations, which are used to show the maximum amount of information in a clear and legible way.

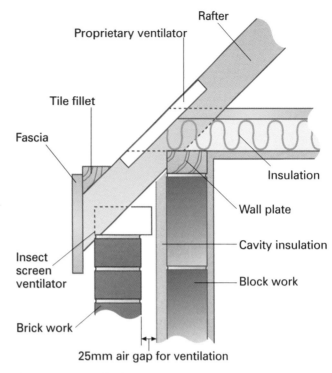

Figure 2.6 Assembly drawing

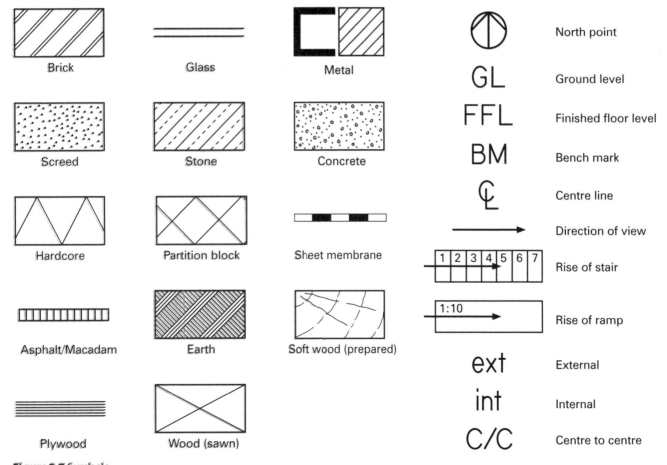

Figure 2.7 Symbols

Item	Abbreviation	Item	Abbreviation
Airbrick	AB	Hardcore	hc
Asbestos	abs	Hardwood	hwd
Bitumen	bit	Insulation	insul
Boarding	bdg	Joist	jst
Brickwork	bwk	Mild steel	MS
Building	bldg	Plasterboard	pbd
Cast iron	CI	Polyvinyl acetate	PVA
Cement	ct	Polyvinyl chloride	PVC
Column	col	Reinforced concrete	RC
Concrete	conc	Satin chrome	SC
Cupboard	cpd	Satin anodised aluminium	SAA
Damp proof course	DPC	Softwood	swd
Damp proof membrane	DPM	Stainless steel	SS
Drawing	dwg	Tongue and groove	T&G
Foundation	fnd	Wrought iron	WI
Hardboard	hdbd		

Table 2.1 Abbreviations

Contract documents

Contract documents are also vital to a construction project. They are created by a team of specialists – the architect, structural engineer, services engineer and quantity surveyor – who first look at the draft of drawings from the architect and client. Just which contract documents this team goes on to produce will vary depending on the size and type of work being done, but will usually include:

- plans and drawings

- specification

- schedules

- bill of quantities

- conditions of contract.

Plans and drawings have already been covered, so here we will start with the specification.

Specification

The specification or 'spec' is a document produced alongside the plans and drawings and is used to show information that cannot be shown on the drawings. Specifications are almost always used, except in the case of very small contracts. A specification should contain:

A good 'spec' helps avoid confusion when dealing with sub-contractors or suppliers

- **site description** – a brief description of the site including the address

- **restrictions** – what restrictions apply such as working hours or limited access

- **services** – what services are available, what services need to be connected and what type of connection should be used

- **materials description** – including type, sizes, quality, moisture content etc.

- **workmanship** – including methods of fixing, quality of work and finish.

The specification may also name sub-contractors or suppliers, or give details such as how the site should be cleared, and so on.

An example of a specification can be found in the fictional project in Chapter 4 (see page 59).

Schedules

A schedule is used to record repeated design information that applies to a range of components or fittings. Schedules are mainly used on bigger sites where there are multiples of several types of house (4-bedroom, 3-bedroom, 3-bedroom with dormers, etc.), each type having different components and fittings. The schedule avoids the wrong component or fitting being put in the wrong house. Schedules can also be used on smaller jobs such as a block of flats with 200 windows, where there are six different types of window.

The need for a specification depends on the complexity of the job and the number of repeated designs that there are. Schedules are mainly used to record repeated design information for:

- doors
- windows
- ironmongery
- joinery fitments
- sanitary components
- heating components and radiators
- kitchens.

A schedule is usually used in conjunction with a range drawing and a floor plan.

The following are basic examples of these documents, using a window as an example:

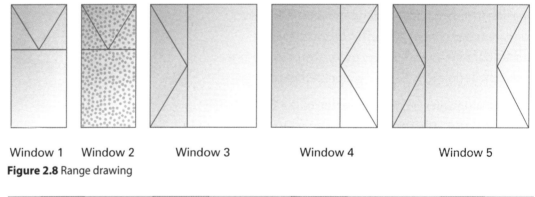

Window 1   Window 2   Window 3   Window 4   Window 5

Figure 2.8 Range drawing

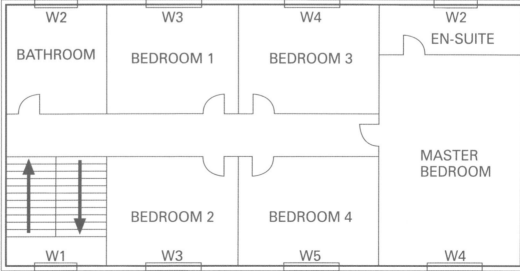

Figure 2.9 Floor plan

WINDOW SCHEDULE		
WINDOW	LOCATIONS	NOTES
Window 1	Stairwell	
Window 2	Bathroom En-suite	Obscure glass
Window 3	Bedroom 1 Bedroom 2	
Window 4	Bedroom 3 Master bedroom	
Window 5	Bedroom 4	

Figure 2.10 Schedule for a window

The schedule shows that there are five types of window, each differing in size and appearance; the range drawing shows what each type of window looks like; and the floor plan shows which window goes where. For example, the bathroom window is a type two window, which is 1200 $\times$ 600 $\times$ 50cm with a top-opening sash and obscure glass.

Bill of quantities

The bill of quantities is produced by the quantity surveyor. It gives a complete description of everything that is required to do the job, including labour, materials and any items or components, drawing on information from the drawings, specification and schedule. The same single bill of quantities is sent out to all **prospective** contractors so they can submit a tender based on the same information – this helps the client select the best contractor for the job.

Every item needed should be listed on the bill of quantities

All bills of quantities contain the following information:

- **preliminaries** – general information such as the names of the client and architect, details of the work and descriptions of the site

- **preambles** – similar to the specification, outlining the quality and description of materials and workmanship, etc.

- **measured quantities** – a description of how each task or material is measured with measurements in metres (linear and square), hours, litres, kilograms or simply the number of components required

- **provisional quantities** – approximate amounts where items or components cannot be measured accurately

- **cost** – the amount of money that will be charged per unit of quantity.

The bill of quantities may also contain:

- any costs that may result from using sub-contractors or specialists

- a sum of money for work that has not been finally detailed

- a sum of money to cover contingencies for unforeseen work.

This is an extract from a bill of quantities that might be sent to prospective contractors, who would then complete the cost section and return it as their tender.

Item ref No	Description	Quantity	Unit	Rate £	Cost £
A1	Treated 50 × 225 mm sawn carcass	200	M		
A2	Treated 75 × 225 mm sawn carcass	50	M		
B1	50 mm galvanised steel joist hangers	20	N/A		
B2	75 mm galvanised steel joist hangers	7	N/A		
C1	Supply and fit the above floor joists as described in the preambles				

Figure 2.11 Sample extract from a bill of quantities

To ensure that all contractors interpret and understand the bill of quantities consistently, the Royal Institution of Chartered Surveyors and the Building Employers' Confederation produce a document called the *Standard Method of Measurement of Building Works* (SMM). This provides a uniform basis for measuring building work, for example stating that carcassing timber is measured by the metre whereas plasterboard is measured in square metres.

Conditions of contract

Almost all building work is carried out under a contract. A small job with a single client (e.g. a loft conversion) will have a basic contract stating that the contractor will do the work to the client's satisfaction, and that the client will pay the contractor the agreed sum of money once the work is finished. Larger contracts with clients such as the Government will have additional clauses, terms or **stipulations**, which may include any of the following.

Variations

A variation is a modification of the original drawing or specification. The architect or client must give the contractor written confirmation of the variation, then the contractor submits a price for the variation to the quantity surveyor (or client, on a small job). Once the price is accepted, the variation work can be completed.

Interim payment

An **interim** payment schedule may be written into the contract, meaning that the client pays for the work in instalments. The client may pay an amount each month, linked to how far the job has progressed, or may make regular payments regardless of how far the job has progressed.

Final payment

Here the client makes a one-off full payment once the job has been completed to the specification. A final payment scheme may also have additional clauses included, such as:

- **retention**
  This is when the client holds a small percentage of the full payment back for a specified period (usually six months). It may take some time for any defects to show, such as cracks in plaster. If the contractor fixes the defects, they will receive the retention payment; if they don't fix them, the retention payment can be used to hire another contractor to do so.

- **penalty clause**
  This is usually introduced in contracts with a tight deadline, where the building must be finished and ready to operate on time. If the project overruns, the client will be unable to trade in the premises and will lose money, so the contractor will have to compensate the client for lost revenue.

Did you know?

On a poorly run contract, a penalty clause can be very costly and could incur a substantial payment. In an extreme case, the contractor may end up making a loss instead of a profit on the project

Find out

For more about the Building Regulations, visit www.ukbuildingstandards.org.uk

Remember

Building regulations can change over time. Always be sure you are using the most updated version.

Building Regulations documentation

Building Regulations are a set of rulings that apply to many construction projects, and are linked with a whole range of documentation that you may come across in your job.

The background to the Building Regulations is covered in detail in Chapter 3, but here we will focus on the actual documents you need to know about.

The regulations are broken down into several categories:

- Part A – Structural safety
- Part B – Fire safety
- Part C – Resistance to moisture and weather
- Part D – Toxic substances
- Part E – Resistance to sound
- Part F – Ventilation
- Part G – Hygiene
- Part H – Drainage and waste disposal
- Part J – Heat-producing appliances
- Part K – Protection from falling
- Part L – Conservation of fuel and power
- Part M – Access to and use of buildings
- Part N – Glazing safety
- Part P – Electrical safety.

Each of these sections contains an 'approved document', detailing what is covered by that part of the regulations:

Approved document A

A1 – Loading

A2 – Ground movement

A3 – Disproportionate collapse

Approved document B

B1 – Means of warning and escape

B2 – Internal fire spread (linings)

B3 – Internal fire spread (structure)

B4 – External fire spread

B5 – Access and facilities for the fire service

Approved document C

C1 – Site preparation and resistance to contaminates

C2 – Resistance to moisture

Approved document D

D1 – Cavity insulation

Approved document E

E1 – Protection against sound from other parts of the building and adjoining buildings

E2 – Protection against sound within a dwelling-house, etc.

E3 – Reverberation in the common internal parts of buildings containing flats or rooms for residential purposes

E4 – Acoustic conditions in schools

Approved document F deals only with ventilation

Approved document G

G1 – Sanitary conveniences and washing facilities

G2 – Bathrooms

G3 – Hot water storage

Approved document H

H1 – Foul water drainage

H2 – Wastewater treatment systems and cesspools

H3 – Rainwater drainage

H4 – Building over sewers

H5 – Separate systems of drainage

H6 – Solid waste storage

Approved document J

J1 – Air supply

J2 – Discharge of products of combustion

J3 – Protection of building

J4 – Provision of information

J5 – Protection of liquid fuel storage systems

J6 – Protection against pollution

Approved document K

K1 – Stairs, ladders and ramps

K2 – Protection from falling

K3 – Vehicle barriers and loading bays

K4 – Protection from collision with open windows, skylights and ventilators

K5 – Protection against impact from and trapping by doors

Approved document L

L1A – Conservation of fuel and power in new dwellings

L1B – Conservation of fuel and power in existing dwellings

L2A – Conservation of fuel and power in new buildings other than dwellings

L2B – Conservation of fuel and power in existing buildings other than dwellings

Approved document M

M1 – Access and use

M2 – Access to extensions to buildings other than dwellings

M3 – Sanitary conveniences in extensions to buildings other than dwellings

M4 – Sanitary conveniences in dwellings

Approved document N

N1 – Protection against impact

N2 – Manifestation of glazing

N3 – Safe opening and closing of windows, skylights and ventilators

N4 – Safe access for cleaning windows, etc.

Approved document P

P1 – Design and installation of electrical installations

Almost all building work requires approval from the Building Regulations Authority. The few exemptions include greenhouses or agricultural buildings, which fall outside parts A–K, M and N, and agricultural buildings, which are also exempt from part P. Some of these may still require planning permission.

General site paperwork

No building site could function properly without a certain amount of paperwork. Here is a brief, but not exhaustive, description of some of the other documents you may encounter. Some companies will have their own forms to cover such things as scaffolding checks.

Timesheet

Timesheets record hours worked, and are completed by every employee individually. Some timesheets are basic, asking just for a brief description of the work done each hour, but some can be complicated. In some cases timesheets may be used to work out how many hours the client will be charged for.

P. Gresford Building Contractors

Timesheet

Employee _____   Project/site _____

Date	Job no.	Start time	Finish time	Total time	Travel time	Expenses
M						
Tu						
W						
Th						
F						
Sa						
Su						
Totals						

Employee's signature _____

Supervisor's signature _____

Date _____

Figure 2.12 Timesheet

Day worksheets

Day worksheets are often confused with timesheets, but are different as they are used when there is no price or estimate for the work, to enable the contractor to charge for the work. Day worksheets record work done, hours worked and sometimes materials used.

P. Gresford Building Contractors

Day worksheet

Customer _Chris MacFarlane_   Date _____

Description of work being carried out _____
Hang internal door in kitchen.

_____

Labour	Craft	Hours	Gross rate	TOTALS

Materials	Quantity	Rate	% addition	

Plant	Hours	Rate	% addition	

Comments

Signed _____   Date _____

Site manager/foreman signature _____

Figure 2.13 Day worksheet

P. Gresford Building Contractors

Job sheet

Customer   Chris MacFarlane

Address   1 High Street
             Any Town
             Any County

Work to be carried out

Hang internal door in kitchen

Special conditions/instructions

Fit with door closer
3 × 75mm butt hinges

Figure 2.14 Job sheet

Job sheet

A job sheet is similar to a day worksheet – it records work done – but is used when the work has already been priced. Job sheets enable the worker to see what needs to be done and the site agent or working foreman to see what has been completed.

VARIATION TO PROPOSED WORKS AT 123 A STREET

REFERENCE NO:

DATE _____

FROM _____

TO _____

POSSIBLE VARIATIONS TO WORK AT 123 A STREET

ADDITIONS

OMISSIONS

SIGNED -------------------------------------

Variation order

This sheet is used by the architect to make any changes to the original plans, including omissions, alterations and extra work.

Figure 2.15 Variation order

CONFIRMATION FOR VARIATION TO PROPOSED WORKS AT 123 A STREET

REFERENCE NO:

DATE _____

FROM _____

TO _____

I CONFIRM THAT I HAVE RECEIVED WRITTEN INSTRUCTIONS FROM _____
POSITION _____
TO CARRY OUT THE FOLLOWING POSSIBLE VARIATIONS TO THE ABOVE NAMED CONTRACT

ADDITIONS

OMISSIONS

SIGNED -------------------------------------

Confirmation notice

This is a sheet given to the contractor to confirm any changes made in the variation order, so that the contractor can go ahead and carry out the work.

Figure 2.16 Confirmation notice

Orders/requisitions

A requisition form or order is used to order materials or components from a supplier.

P. Gresford Building Contractors

Requisition form

Supplier _____   Order no. _____

_____   Serial no. _____

Tel no. _____   Contact _____

Fax no. _____   Our ref _____

Contract/Delivery address/Invoice address   Statements/applications

_____   for payments to be sent to

_____   _____

Tel no. _____   _____

Fax no. _____   _____

Item no.	Quantity	Unit	Description	Unit price	Amount

Total £ _____

Payment terms _____   Date _____

Originated by _____

Authorised by _____

Figure 2.17 Requisition form

Delivery notes

Delivery notes are given to the contractor by the supplier, and list all the materials and components being delivered. Each delivery note should be checked for accuracy against the order (to ensure what is being delivered is what was asked for) and against the delivery itself (to make sure that the delivery matches the delivery note). If there are any **discrepancies** or if the delivery is of a poor quality or damaged, you must write on the delivery note what is wrong *before* signing it and ensure the site agent is informed so that he/she can rectify the problem.

Bailey & Sons Ltd

Building materials supplier

Tel: 01234 567890

Your ref: AB00671

Our ref: CT020   **Date:** 17 Jul 2006

Order no: 67440387

Invoice address: **Delivery address:**
Carillion Training Centre,   Same as invoice
Deptford Terrace, Sunderland

Description of goods	Quantity	Catalogue no.
OPC 25kg	10	OPC1.1

Comments:
Date and time of receiving goods:
Name of recipient (caps):
Signature:

Figure 2.18 Delivery note

Remember

Invoices may need paying by a certain date – fines for late payment can sometimes be incurred – so it is important that they are passed on to the finance office or financial controller promptly

Invoices

Invoices come from a variety of sources such as suppliers or sub-contractors, and state what has been provided and how much the contractor will be charged for it.

INVOICE

JARVIS BUILDING SUPPLIES
3rd AVENUE
THOMASTOWN

L Weeks Builders
4th Grove
Thomastown

Quantity	Description	Unit price	Vat rate	Total
30	Galvanised joint hangers	£1.32	17.5%	£46.53
			TOTAL	£46.53

To be paid within 30 days from receipt of this invoice

Please direct any queries to 01234 56789

Figure 2.19 Invoice

Delivery records

Delivery records list all deliveries over a certain period (usually a month), and are sent to the contractor's Head Office so that payment can be made.

JARVIS BUILDING SUPPLIES
3rd AVENUE
THOMASTOWN

Customer ref_____

Customer order date_____

Delivery date_____

Item no	Qty Supplied	Qty to follow	Description	Unit price
1	30	0	Galvanised joinst hangers	£1.32

Delivered to: L Weeks builders
4th Grove
Thomastown
Customer signature _____

Figure 2.20 Delivery record

DAILY REPORT/SITE DIARY

PROJECT_____
DATE_____

Identify any of the following factors, which are affecting or may affect the daily work activities and give a brief description in the box provided

WEATHER ( )   ACCESS ( )   ACCIDENTS ( )   SERVICES ( )
DELIVERIES ( )   SUPPLIES ( )   LABOUR ( )   OTHER ( )

SIGNED _____
POSITION _____

Figure 2.21 Daily report or site diary

Daily report/site diary

This is used to pass general information (deliveries, attendance etc.) on to a company's Head Office.

Remember

Remember – you should always check a delivery note against the order and the delivery itself, then write any discrepancies or problems on the delivery note *before* signing it

Accident and near miss reports

It is a legal requirement that a company has an accident book, in which reports of all accidents must be made. Reports must also be made when an accident nearly happened, but did not in the end occur – known as a 'near miss'. It is everyone's responsibility to complete the accident book. If you are also in a supervisory position you will have the responsibility to ensure all requirements for accident reporting are met.

Safety tip

If you are involved in or witness an accident or near miss, make sure it is entered in the book – for your own safety and that of others on the site. If you don't report it, it's more likely to happen again

---

Report of an Accident, Dangerous Occurrence or Near Miss

Date of incident _____   **Time of incident** _____

Location of incident _____

Details of person involved in accident

Name _____   Date of birth _____

Address _____

_____   Occupation _____

Date off work (if applicable) _____   **Date returning to work** _____

Nature of injury _____

Management of injury   ☐ First Aid only   ☐ Advised to see doctor

☐ Sent to casualty   ☐ Admitted to hospital

Account of accident, dangerous occurrence or near miss
(Continued on separate sheet if necessary)

Witnesses to the incident
(Names, addresses and occupations)

Was the injured person wearing PPE? If yes, what PPE? _____

_____

Signature of person completing form _____

Occupation _____   **Date** _____

Figure 2.22 Accident/near miss report

On the job: Complying with approvals

Brian and Amanda have been approached by a friend to do a loft conversion. They apply for Planning and Building Regulations approval and are given both, so they carry out the work. They also come across a problem with the chimney and decide to remove some of the bricks. With the work completed, the Building Inspector shows up to check the job. What can the Inspector do? What effect will this have on the job? What could have been done to prevent it?

FAQ

How do I know what scale the drawing is at?

The scale should be written on the title panel (the box included on a plan or drawing giving basic information such as who drew it, how to contact them, the date and the scale).

How do I know if I need a schedule?

Schedules are only really used in large jobs where there is a lot of repeated design information. If your job has a lot of doors, windows etc., it is a good idea to use one.

How do I know if I need approval?

If you are unsure, check section three of the Building Regulations or contact your local authority.

Do I need to know all the different Building Regulations and what is contained in each section?

No, but a good understanding of what is involved is needed.

How many different forms are there?

A lot of forms are used and some companies use more than others. You should ensure you get the relevant training on completing the form before using it.

Knowledge check

1. Who draws the plans?

2. State three different types of drawings and give a suitable scale for each one.

3. State three of the main contract documents.

4. What is the main purpose of a specification?

5. What is the purpose of the bill of quantities?

6. What is a penalty clause?

7. What does the approved document A cover?

8. Which approved document deals with stairs?

9. What is the role of the Building Inspector?

chapter 3

Planning and work programmes

OVERVIEW

Any building project begins long before the first brick is laid or the first foundation dug. Most buildings and construction projects will need some sort of planning approval before they get underway, as a range of planning restrictions are in place to keep building standards up, protect local people and protect the environment.

Work planning is also of paramount importance for every job, whether a single dwelling or a large housing estate. Without it even the smallest job can go wrong: something simple is forgotten or omitted, such as ordering a skip, and the job is suddenly delayed by anything up to a week. On a smaller job, poor planning can result in delays, which will harm your reputation and jeopardise future contracts. With larger contracts, penalty clauses can be costly: if the job overruns and isn't finished on time, the client may claim substantial amounts of money from the contractor. This chapter will deal with:

- external planning restrictions
- work programming.

External planning restrictions

Before starting to plan a building project, it is important to know how your plans may be affected by local and national building restrictions. The two main sets of restrictions you will come across are:

- the Building Regulations
- planning permission.

It is crucial that anyone planning a construction project understands how these work, and seeks the necessary approval in the correct way. If not, building work runs the risk of having to be halted, altered or even taken down.

The Building Regulations

The Building Regulations were first introduced in the late 1800s to improve the appalling housing conditions common then. The Public Health Act 1875 allowed local authorities to make their own laws regarding the planning and construction of buildings. There were many grey areas and **inconsistencies** between local authorities, especially where one authority bordered another.

Building Regulations help protect the environment

This system remained in place for almost a century until the Building Regulations 1965 came into force. These replaced all local laws with a uniform Act for all in England and Wales to follow. The only exception was inner London, which was covered by the London Building Acts. The Government passed a new law in 1984, setting up the Building Regulations 1985 to cover all England and Wales, including inner London.

The current law is the Building Regulations 2000, amended in April 2006 to take into account things such as wheelchair access and more environmentally friendly practices. The current law also covers all England and Wales.

Scotland is governed slightly differently and is covered by the Building (Scotland) Act 2003. Northern Ireland is covered by the Building (Amendment) Regulations (Northern Ireland) 2006 which came into effect on November 2006.

The main purpose of the Building Regulations is to ensure the health, safety and welfare of all people in and around buildings as well as to further energy conservation and to protect the environment. The regulations apply to most new buildings as well as any alterations to existing buildings, whether they are domestic, commercial or industrial. Many projects also require planning permission, which is covered on pages 43–45.

These are the types of work classified as needing Building Regulations approval:

- the erection of an extension or building

- the installation or extension of a service or fitting which is controlled under the regulations

- an alteration project involving work which will temporarily or permanently affect the ongoing compliance of the building, service, or fitting with the requirements relating to structure, fire, or access to and the use of the building

- the insertion of insulation into a cavity wall

- the underpinning of the foundations of a building

- work affecting the thermal elements, energy status or energy performance of the building.

If you are unsure whether the work you are going to carry out needs Building Regulations approval, contact the local authority.

The Building Regulations are enforced by two types of building control bodies: local authority building control and Approved Inspector building control. If you wish to apply for approval, you must contact one of these bodies.

If you use an Approved Inspector, you must contact the local authority to tell them what is being done where, stating that the Inspector will be responsible for the control of the work.

The Building Inspector will need to be involved at every key stage

If you choose to go to the local authority, there are three ways of applying for consent:

- **full plans** – Plans are submitted to the local authority along with any specifications and other contract documents. The local authority scrutinises these and makes a decision.

- **building notice** – A less detailed amount of information is submitted (but more can be requested) and no decision is made. The approval process is determined by the stage the work is at.

- **regularisation** – This is a means of applying for approval for work that has already been completed without approval.

The Building Inspector will make regular visits to ensure that the work is being carried out to the standards set down in the application, and that no extra unapproved work is being done. Often the contractor will tell the Inspector when the job has reached a certain stage, so that they can come in and check what has been done. If the Inspector is not informed at key stages, he/she can ask for the work to be opened up to be checked.

Building Regulations approval is not always given but there is an appeals procedure. For more information, contact your local authority.

Planning permission

As has already been mentioned, with most contracts you must have planning permission as well as Building Regulations approval before starting the work.

Planning permission laws were introduced to stop people building whatever they like wherever they like. The submission of a planning application gives both the local authority and the general public a chance to look at the development, to see if it is in keeping with the local area and whether it serves the interests of the local community.

The main **remit** of planning laws is to control the use and development of land in order to obtain the greatest possible environmental advantages with the least inconvenience for both the person/s applying for permission and society as a whole.

The key word in planning is 'development', defined in planning law as 'the carrying out of building, engineering, mining or other operations in, on, over or under land, or the making of any material change in the use of any buildings or other land'. As well as building work, this covers the construction of a new road or driveway, and even change of use: if a bank is to be turned into a wine bar, planning permission will be needed.

Planning permission is required for most forms of development. Here are a few more examples of work requiring planning permission:

- virtually all new building work
- house extensions including conservatories, loft conversions and roof additions (such as dormers)
- buildings and other structures on the land including garages
- adding a porch to your house
- putting up a TV satellite dish.

Did you know?

Planning permission is needed if you want to put up a satellite dish. The job itself is small and not disruptive, but a dish is thought to change the outer appearance of a house enough to need permission

The public has a right to know about proposed developments

Even if you are intending to work from home and wish to convert part of your home into an office, you will require planning permission if:

- your home is no longer to be used mainly as a private residence
- your business creates more traffic or creates problems with parking due to people calling
- your business involves any activities classed as unusual in a **residential** area
- Your business disturbs your neighbours at unreasonable hours or creates other forms of nuisance or smell.

Not all work requires planning. You can make certain types of minor alterations to your house, such as putting up a fence or dividing wall (providing it is less than 1 metre high next to a highway, or under 2 metres elsewhere), without planning permission.

In areas such as conservation areas or classified Areas of Outstanding Natural Beauty there will be stricter controls on what is allowed. Listed buildings also have stricter controls and come under the Planning (Listed Buildings and Conservation Areas) Act 1990.

For planning permission, you must apply to your local council. When they look at your proposed works, they will take into consideration:

- the number, size, positioning, layout and external appearance of the buildings
- the proposed means of access, landscaping and impact on the neighbourhood
- **sustainability**, and whether the necessary infrastructure, such as roads, services, etc., will be available
- the proposed use of the development.

Several steps are involved in applying for planning permission. The first is to contact the local council to see if they think planning permission is required (some councils may charge a small fee for this advice). If they say you do need planning permission, you need to then ask them for an application form. There are two types of planning permission that you can apply for:

- **outline application** This can be made if you want to see what the council thinks of the building work you intend to do before you go to the trouble of having costly plans drawn up. Details of the work will have to be submitted later if the outline application is successful.
- **full application** Here a full application is made with all the plans, specifications, and so on.

Once you have completed the relevant form this must be sent to the council along with any fee.

Next, the contents of your application will be publicised so that people can express their views and raise any objections. A copy will be placed in the planning register; an electronic version will be placed on the council's website; and immediate neighbours will be written to (or a fixed

Find out

Look online for your local authority's website. Most now have the main information you need for planning matters, and you may even be able to download the forms you need to use notice will be displayed on or as near as possible to the site). The council may also advertise your application in a local newspaper. As the applicant, you will be entitled to have a copy of any reports, objections and expressions of support the council receives regarding your application.

The council normally takes up to eight weeks to make a decision on your application but in some cases it may take longer. If this happens, the council should write asking for your written consent to extend the period. If your application is not dealt with within eight weeks, you can appeal to the Secretary of State, but this can be a lengthy procedure itself, so it is best to try to resolve the matter at a local level.

In looking at an application, the council considers whether there are valid reasons for refusing or granting permission: the council cannot simply reject a proposal because many people oppose it. The council will look at whether your proposal is consistent with the area's appearance, whether it will cause traffic problems and whether it has any impact on local amenities, environment and services.

Once an application has been looked at, there are four possible outcomes: permission refused; application still pending; granted with conditions; or granted.

- **Permission refused**

    If permission is refused, the council must state its reasons for turning down the application. If you feel these are unfair, you can appeal to the Secretary of State. Appeals must be made within six months of the council's decision and are intended as a last resort. It can take months to get a decision, which may be a refusal. Alternatively, you can ask what changes need to be made to allow the proposal to pass: if these are acceptable, the amended application can be submitted for processing. If after this the application is still rejected, the work cannot go ahead. However, different authorities have different procedures, so always check before submitting proposals.

- **Application still pending**

    Here the council may have found that it needs extra time to allow comments to come in, or to deal with particular issues that have arisen. If the application is still pending then, as stated previously, the council must ask for your written consent to extend the period for making a decision.

- **Granted with conditions**

    In this case you are able to start the work, remembering to comply with the conditions stated. If you fail to comply, permission will be revoked and you may be ordered to undo the work done. If you are unhappy with the conditions set, you can ask for advice and, if needs be, make alterations to the plans. This would mean resubmitting the application.

- **Granted**

    If you have been granted permission, you are free to start the work.

Remember

If you build something without planning permission then you may be forced to **dismantle** the building and put it back to the original state – as well as paying for the work yourself!

Work programming

Once planning permission and Building Regulations approval have been obtained, the next step is to plan the work (NB in some instances the client may ask the contractor to provide a work programme at the tender stage, to check the contractor's efficiency and organising ability).

A work programme is vital for good work planning, as it shows:

- what tasks are to be done and when, including any overlap in the tasks
- what materials are required and when
- what plant is needed, when and for how long
- what type of workforce is required and when.

A few different types of work programme are in use, and we will cover the main two on pages 48–52.

Planning the site

For every fair-sized job, the building site needs to be carefully planned. A poorly planned site can cause problems and delays, as well as incurring costs and even causing accidents.

A building site should be seen as a temporary workshop, store and office for the contractor, and must contain all the **amenities** needed on a permanent base. Sites should be planned in a way that minimises the movement of employees, materials and plant throughout the construction, while at the same time providing protection and security for employees, materials and components, and members of the public. A well-planned site will also have good transport routes, which will not disrupt the site or the general traffic.

Many things need to be included on a building site, so it is often easiest to plan your site using a site plan and cut-outs of the amenities you need. These cut-outs can be laid onto the plan and moved around until a suitable layout is found.

The ideal layout of the site will vary according to the size and **duration** of the job – there is no point hiring site offices for a job that will only last a day! The following gives an idea of what might be needed on an average site:

- **site offices**

  The office space (usually portable cabins) should be of a decent size, usually with more than one room for different members of staff and a large room for meetings. Phone, fax and email facilities will be needed, so that the site office can communicate with Head Office, contractors, suppliers and others. As with any office, the site office must be heated, have plenty of light (natural or artificial) and be fitted out with useful, comfortable furniture.

Remember

If you need to plan several sites, save the cut-outs from one to use on the next (checking that you are using the same scale). You could end up with a 'kit' to use whenever you need it.

- **first aid office**

  This is sometimes contained within the site office, but on larger sites a separate space may be needed so that injured people can be treated quickly and efficiently. The first aid office must be fully stocked, and there must be sufficient trained first aiders on site.

- **toilets**

  There must be sufficient toilets on the site. Usually there will be a WC block next to the canteen or mess area, with additional portable toilets dotted around the site if needed. Toilets must be kept clean and well stocked at all times, and have somewhere for people to wash their hands. The WC block may also need to house showers if the work being done requires them.

- **lunch area**

  This should be protected from the wind and rain and have heating and electricity. It should contain equipment such as a microwave, kettle or urn and fridge to heat and keep food, as well as suitable food storage such as cupboards. There should be adequate seating and tables, and the space should be kept clean to prevent any unwelcome pests such as rats or cockroaches.

- **drying room**

  This provides space for employees to dry off any clothes that get wet, on the way to or during work. It is usually sited next to the lunch area, or is part of the same building. The room must have adequate heating and ventilation, as well as lockers or storage to house things like motorcycle helmets.

- **cranes, hoists, etc.**

  These can be static or portable. When a large static crane is required, its position needs to be planned so that it can easily and safely reach the area where it is needed. Larger cranes should be situated away from the main site office for safety reasons.

- **transport route**

  Having a good transport route into, out of and within a building site is vital. It is best to have separate entrances and exits, with a one-way system on the site and good signposting throughout. These measures will avoid large delivery lorries having to turn around on site, and help to keep both internal and external traffic flowing with minimum disruption.

- **waste area**

  This must be well away from the lunch area for health and safety reasons, and should be easily accessible from the transport route so that the skips and bins can be emptied easily. Separate well-labelled skips are needed for different kinds of refuse, and there should be some for recycling. Certain skips should be kept separate to avoid **contamination**, and chemical dumps (for paint, etc.) should be kept secure and emptied regularly.

Various types of storage are also needed on a building site, such as:

- **materials storage** – enough adequate space to store all types of materials, ideally near to where they are being used (for example, cement and sand should be stored near the mixer). All materials should be stored in a way that prevents them being damaged or stolen; some materials will have to be stored separately to avoid contamination.

- **component storage** – a secure compound protected from the wind and rain for items such as doors and windows. Again, components should be stored in a way that prevents them being damaged.

- **tool storage** – a secure place for employees' own tools as well as site tools such as table saws. The tool storage area needs to be thoroughly secure to prevent theft.

- **ironmongery storage** – a locked compound in a container with well-labelled racks to avoid things like screws and nails being mixed up. Expensive ironmongery such as door furniture needs to be properly secure. On a well-planned site, expensive ironmongery is only ordered when needed.

A good site layout might look something like this.

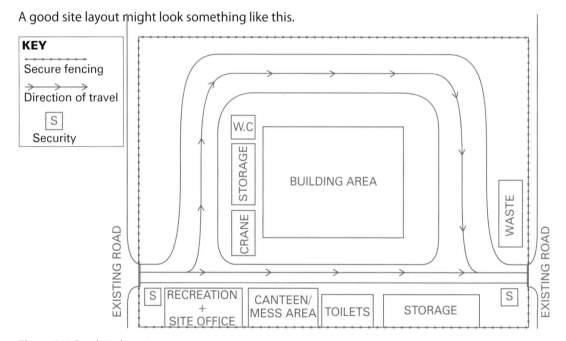

Figure 3.1 Good site layout

Planning the work

There are many types of work programme, including the critical path and the Bar/Gantt chart. The latter is the one you will come across most often.

Bar charts

The bar or Gantt chart is the most popular work programme as it is simple to construct and easy to understand. Bar charts have tasks listed in a vertical column on the left and a horizontal timescale running along the top.

Time in days										
Activity	1	2	3	4	5	6	7	8	9	10
Dig for foundation and service routes										
Lay foundations										
Run cabling, piping etc. to meet existing services										
Build up to DPC										
Lay concrete floor										

Figure 3.2 Basic bar chart

Each task is given a proposed time, which is shaded in along the horizontal timescale. Timescales often overlap as one task often overlaps another.

Time in days										
Activity	1	2	3	4	5	6	7	8	9	10
Dig for foundation and service routes	■	■								
Lay foundations			■	■						
Run cabling, piping etc. to meet existing services				■	■					
Build up to DPC						■	■			
Lay concrete floor								■	■	■

Key: proposed ■   actual ■

Figure 3.3 Bar chart showing proposed time for a contract

The bar chart can then be used to check progress. Often the actual time taken for a task is shaded in underneath the proposed time (in a different way or colour to avoid confusion). This shows how what *has* been done matches up to what *should have* been done.

Time in days										
Activity	1	2	3	4	5	6	7	8	9	10
Dig for foundation and service routes	■	■								
Lay foundations			■	■						
Run cabling, piping etc. to meet existing services				■	■					
Build up to DPC							■	■		
Lay concrete floor									■	■

Figure 3.4 Bar chart showing actual time half way through a contract   Key: proposed ■   actual ■

As you can see, a bar chart can help you plan when to order materials or plant, see what trade is due in and when, and so on. A bar chart can also tell you if you are behind on a job; if you have a penalty clause written into your contract, this information is vital.

When creating a bar chart, you should build in some extra time to allow for things such as bad weather, labour shortages, delivery problems or illness. It is also advisable to have contingency plans to help solve or avoid problems, such as:

- capacity to work overtime to catch up time

- bonus scheme to increase productivity

- penalty clause on suppliers to try to avoid late or poor deliveries

- source of extra labour (e.g. from another site) if needed.

Good planning, with contingency plans in place, should allow a job to run smoothly and finish on time, leading to the contractor making a profit.

Critical paths

Another form of work programme is the critical path. Critical paths are rarely used these days as they can be difficult to decipher. The final part of this chapter will give a brief overview of the basics of a critical path, in case you should come across one.

A critical path can be used in the same way as a bar chart to show what needs to be done and in what sequence. It also shows a timescale but in a different way to a bar chart: each timescale shows both the minimum and the maximum amount of time a task might take.

The critical path is shown as a series of circles called event nodes. Each node is split into three: the top third shows the event number, the bottom left shows the earliest start time, and the bottom right the latest start time.

Did you know?

Bad weather is the main external factor responsible for delays on building sites in the UK. A Met Office survey showed that the average UK construction company experiences problems caused by the weather 26 times a year

The nodes are joined together by lines, which represent the tasks being carried out between those nodes. The length of each task is shown by the times written in the lower parts of the nodes. Some critical paths have information on each task written underneath the lines that join the nodes, making them easier to read.

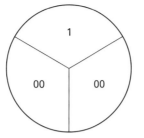

Figure 3.5 Single event node

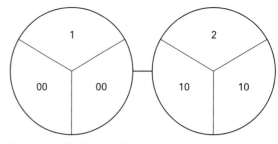

Figure 3.6 Nodes joined together

On a job, many tasks can be worked on at the same time, e.g. the electricians may be wiring at the same time as the plumber putting in his pipes. To show this on a critical path, the path can be split.

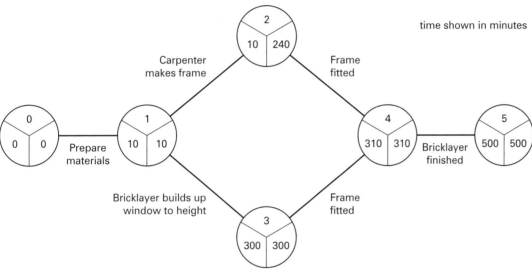

Figure 3.7 Split path

The example shown shows how a critical path can be used for planning building in a window opening, with a carpenter creating a dummy frame.

The event nodes work as follows:

- **Node 0 –** This is the starting point.

- **Node 1 –** This is the first task, where the materials are prepared.

- **Node 2 –** This is where the carpenter makes the dummy frame for the opening. Notice that the earliest start time is 10 minutes and the last start time is 240 minutes. This means that the carpenter can start building the frame at any time between 10 minutes and 240 minutes into the project. This is because the frame will not be needed until 300 minutes, but the job will only take 60 minutes. If the carpenter starts *after* 240 minutes, there is a possibility that the job may run behind.

- **Node 3 –** This is where the bricklayer must be at the site, ready for the frame to be fitted at 300 minutes, or the job will run behind.

- **Node 4** – With the frame fitted, the bricklayer starts at 310 minutes and has until node 5 (500 minutes) to finish.

- **Node 5** – The job should be completed.

When working with a split path it is vital to remember that certain tasks have to be completed before others can begin. If this is not taken into account on the critical path, the job will run over (which may prove costly, both through penalty clauses and also in terms of the contractor's reputation).

On a large job, it can be easy to misread a critical path as there may be several splits, which could lead to confusion.

Remember

Whichever way you choose to programme your work, your programme must be realistic, with clear objectives and achievable goals

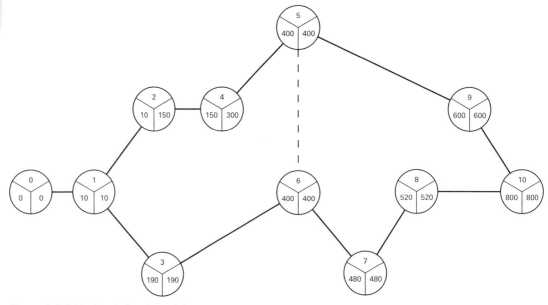

Figure 3.8 Critical path for a large job

On the job: Planning a site

Sanjit is planning a site. When he sets out what is needed with regards to toilets, lunch areas and so on, he opts to use the same entrance and exit for site transport. What problems can this cause, both on and off site?

FAQ

How do I know if my job needs planning permission?

If you are unsure, you should contact your local council.

What type of planning permission should I apply for?

If you are unsure of your work, you can make an outline application, which will tell you if your job will pass without getting costly plans made up (though you will have to submit plans later). If you are confident of what you want, you can apply for a full application.

How much does planning permission cost?

The costs vary depending on what application you make and to which council you make it.

Do I need to have all the listed amenities on my building site?

No. The amenities listed are a guide to what should be on a large site. If you are just doing an extension, the amenities needed will be fewer and simpler (e.g. no site office).

Which type of programme should I use: bar chart or critical path?

It is up to the individual which programme they use – both have their good points – but a bar chart is the easiest to set up and work from.

Knowledge check

1. When were the first Building Regulations covering all England and Wales introduced?

2. Why were planning laws introduced?

3. What do planning laws define as 'development'?

4. If you plan to use your home as an office, what changes require planning permission?

5. State the two types of planning you can apply for.

6. What are the four possible outcomes from the council regarding a planning application?

7. List four things that might be included in the layout for a large site.

8. State four pieces of information you can get from a bar chart.

9. What is the purpose of a contingency plan?

10. With regard to critical paths, what three things are contained in an event node?

chapter 4

The building process

OVERVIEW

The building process covers the construction of a building from start to finish. In this chapter we will look at what and who is involved at every step. We will follow a fictitious job through all stages of the building process showing what documents are used when, as well as who is responsible for carrying out the work and when.

This chapter is meant as a guide only. In some circumstances, or in certain areas of the country, the process may differ slightly. The documents used in our example may not all be needed for every job, but an awareness of each document is very important.

The information is broken down into several sections for easy reference, as follows:

- who's who
- getting started
- contract documents
- specification
- bill of quantities
- Building Regulations and planning permission
- work programming
- site set-up
- starting the work
- completing the work.

Who's who

Many different people are involved in the process of building, and together they are known as the building team. For a full rundown of members of the building team, you can refer to *Carpentry and Joinery NVQ and Technical Certificate Level 2* pages 6–9, but here is a brief description of the main players:

Client – every single job starts with the client, the most important person. Without the client, there is no job.

Architect – works closely with the client and produces contract documents such as the plans and specification.

Local authority – is responsible for checking whether the construction meets all planning and Building Regulations.

Health and Safety Inspectors – are employed by the Health and Safety Executive to ensure that all building work is done in line with health and safety regulations.

Quantity surveyor – works with the architect and client to produce the bill of quantities, which is sent to contractors to enable them to submit tenders.

Contractor – works for the client and carries out the building work in line with the plans.

Sub-contractor – is employed by the contractor to carry out specialist work that the contractor cannot do themselves.

Site agent – is responsible for the day-to-day running of the site, including monitoring the programme of work.

Foreman – works under the site agent and is responsible for organising the work of the craft operatives and sub-contractors.

Operatives – the people, skilled or semi-skilled, who actually carry out the work.

Remember

Health and safety during the course of a contract is paramount and health and safety legislation applies all the way through the contract, from start to finish. Throughout this chapter we will remind you of what health and safety documents, contacts or actions need to be completed

Getting started

Every building contract begins with a client who needs work done, whether it is an extension or a whole scheme of houses. The example we will use in this chapter is the building of a garage with an upstairs office. There are normally two ways the client can proceed:

1. The client contacts contractors who will come out, have a look at the job and price it up. The client picks the contractor and agrees the price, then the contractor does the work. With this method only some of the contract documents are used. The client will be left to apply for planning and Building Regulations approval without the aid of an architect.

2. The client contacts an architect to discuss the work and from then on the architect takes control, acting as the client's representative. With this method, all of the contract documents (including the bill of quantities) are used, and the job's progress is recorded on a programme.

It is possible to mix the two methods, but generally method 1 is used on small jobs such as extensions or loft conversions while method 2 is used for larger jobs such as new housing estates or supermarkets. Method 2 is more expensive as it uses more members of the building team, but this approach ensures a properly planned job.

Please note: method 1 would be the one normally used for our example (the building of a garage with an upstairs office) but for the purposes of this book and to show you more of the building process, we will imagine that the client has chosen to use method 2.

First the client contacts an architect and arranges a meeting, at which the client tells the architect their wishes. The architect then does rough sketches of what the client wants and prepares a rough specification. If the client is happy with the architect's work, he/she will give the architect the go ahead to prepare full contract documents.

Contract documents

The architect starts with the plans, doing several different plans including block plans, site plans, general location drawings and detailed drawings (examples of all of these can be found in Chapter 2, pages 18–22). Below are the types of the drawings that would be required, to give you an idea of the job and what it entails. Full architects' drawings would give more information, such as the positioning of services.

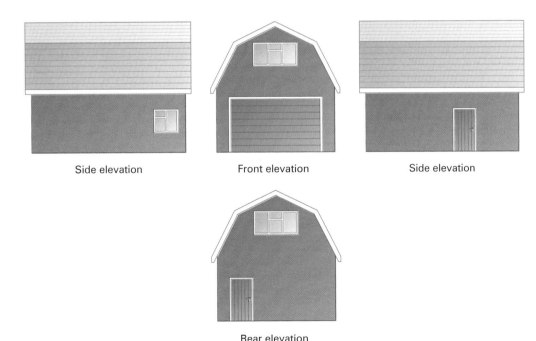

Side elevation   Front elevation   Side elevation

Rear elevation

Figure 4.1 Elevation drawings

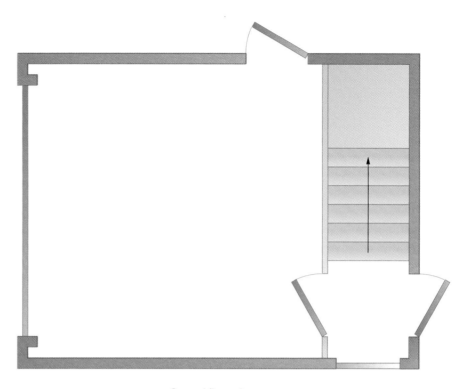

Ground floor plan

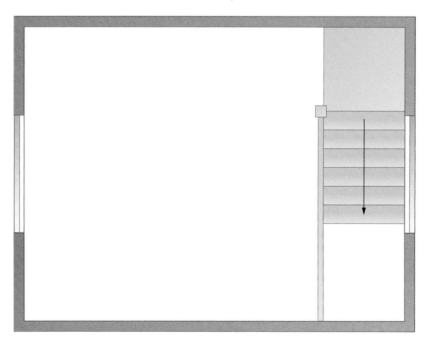

First floor plan

Figure 4.2 Floor plans

Once all the drawings are completed and the client is happy with them, the architect then draws up a specification and, if needed, a schedule. A schedule is not required for this project but an example schedule can be found on page 25.

Specification

The specification is next. This is used alongside the main drawings and gives information that cannot be worked out from the drawings. A full specification for even a small job like this can run to several pages, so what you see below is only a brief extract, covering the workmanship and materials used for constructing the upper floor.

Joists – 47 × 220mm pre-cut, pressure **impregnated**, structurally graded joists @ max. 400mm centres. Joists must be **regularised** and laid with the rounds facing up

Joist hangers – galvanised steel joist hangers built into brickwork with associated fixings for securing joists in place

Insulation – 100mm foil-backed, rigid insulation and 100mm high density (24 kg/m³) Rockwool insulation fitted between joists + 400mm wide strip of 100mm insulation around perimeter

Flooring – 22mm tongued and grooved, moisture resistant flooring grade chipboard (V313 – TG4). Flooring to be glued at every joint and screwed @ no less than 8" centres

Solid blocking, noggins and herringbone strutting, as appropriate

Figure 4.3 Flooring specification

Definition

Regularised joists – joists run through a saw to ensure that they are all the same depth, which will ensure a flat, even ceiling and floor

The next two sections are very closely tied together. Either can be done first but to save time it is ideal to do both at the same time.

Bill of quantities

The quantity surveyor uses all the contract documents from the architect to help draw up the bill of quantities.

Again, even for such a small job the bill of quantities can run to several pages, so for our example we will look at a bill of quantities for just the upper floor part of the contract.

Item ref no	Description	Quantity	Unit	Rate £	Cost £
UF 1	47 × 220mm pre-cut, pressure impregnated, structurally graded joists	180	M	2.79	502.20
UF 2	50mm galvanised steel joist hangers	54	N/A	1.00	54.00
UF 3	100mm foil-backed, rigid insulation	10	N/A	22.78	227.80
UF 4	100mm high density (24kg/m^3) Rockwool insulation	10	N/A	15.68	156.80
UF 5	22mm tongued and grooved, moisture resistant flooring grade chipboard (V313 – TG4)	50	N/A	5.46	273.00
UF 6	Fixings and adhesives	N/A	N/A	50.00	50.00
UF 7	Labour to fit the above to the required specification	64	N/A	15.00	960.00
			Total		2223.80

Figure 4.4 Bill of quantities

Building Regulations and planning permission

The contract documents are sent to the local council for planning permission approval, which could take up to eight weeks to come through. The plans are also sent to the local authority for Building Regulations approval. At this stage it can be helpful to contact the local Health and Safety Executive outlining what is being done. This gives them the opportunity to raise any immediate concerns they may have with the contract or to give any advice. It is also advisable to complete a F10 form for the Health and Safety Executive, informing them that construction work is about to take place. The F10 is only needed if the contract will last longer than 30 calendar days or 500 people days.

Definition

People days – a way of expressing how long it will take to do something, by looking at how many people will be needed for how long. For example, if you need four people, each working for two days, your total number of people days is 4 x 2 = 8

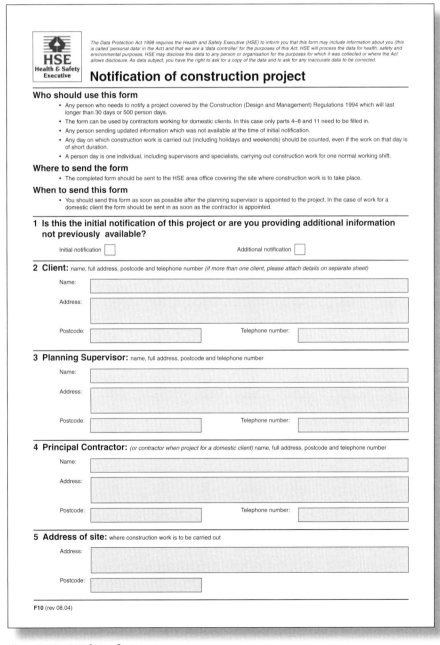

Figure 4.5a F10 form, first page

Did you know?

The cheapest tender is not always the one that wins the contract. Some contractors include in their tender bids that they will recycle so much waste, or employ so many people from the local workforce. Things like this can be the deciding factors in winning a contract

6 Local Authority:
Name of the local government council or authority within whose area the construction work is to be carried out.

7 Please give your estimates on the following:
Please indicate if these estimates are: original ☐   revised ☐   (tick relevant box)

a. The planned date for the commencement of the construction work

b. How long the construction work is expected to take (in weeks)

c. The maximum number of people carrying out construction work on site at any one time

d. The number of contractors expected to work on site

8 Construction work: give brief details of the type of construction work that will be carried out

9 Contractors: name, full address and postcode of those who have been chosen to work on this project (if required continue on a separate sheet). (Note this information is only required when it is known at the time notification is first made to HSE. An update is not required)

10 Declaration of planning supervisor
I hereby declare that   (name of organisation)   has been appointed as planning
                                                                                   supervisor for the project

Signed by or on behalf
of the organisation

Print name   Date

11 Declaration of principal contractor
I hereby declare that   (name of principal contractor)   has been appointed as principal
                                                                                           contractor for the project

Signed by or on behalf
of the organisation

Print name   Date

Figure 4.5b F10 form, second page

Once the planning permission and Building Regulations approvals are received, it is time to select a contractor. At this stage any clauses such as penalty clauses are introduced – the contractor must be made aware of all such clauses before signing the contract. Once the contractor has been selected from the list of tenders, it is time to start planning the work programme.

The next two sections should also be done at the same time. To save time they can sometimes be done earlier than this but, if the original plans have to be changed to fit with Building Regulations or planning permission, this may prove costly.

Work programming

The bar chart for our example looks something like this:

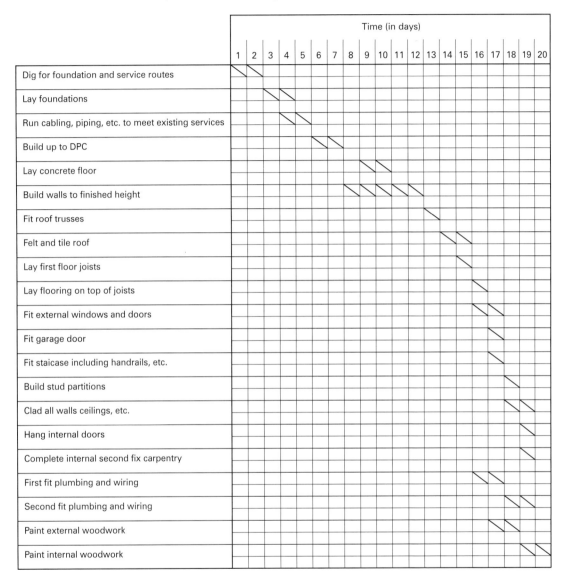

Key: proposed ◢   actual ◩

Figure 4.6 Work programme

As you can see the work programme is divided into two with the 'proposed' section completed. The 'actual' section will be completed as the job progresses.

Once the programme has been set up and start dates have been confirmed, it is time for the contractor to contact suppliers, sub-contractors, etc., and to order materials, labour and plant.

You can see from the programme that the first task is to get the foundations done, so it is vital that all the relevant plant, labour and materials are on site ready to start on the required date. Carpenters do not need to be on site yet, but it is good practice to ensure in advance that they will be available to start work on the required dates.

It is bad practice to order all the materials this early (having items like doors delivered to site at this stage could lead to them being damaged) but the supplier must be aware of what materials will be needed later and when. Good forward planning can make the difference between having a well-run project with a profit and a badly run project with a loss.

Now contingency plans need to be set up, to help if the work runs behind due to poor weather, etc. As our example is not a large job, a contingency plan is not absolutely vital, but it is always advisable to have some sort of plan in place. Our job has a planned duration of only 20 days, so it is not feasible to pull contractors in from other sites. The main contingency here is to offer cash bonus incentives and overtime payments so that the work can be pulled back on track if it falls behind.

Site set-up

Although this is not a large site, and not all the amenities will be needed, there should still be a site layout plan showing where materials will be stored, and so on. A full plan for a site layout is on page 48.

Starting the work

If the planning stage has been done successfully, the materials should arrive on site a few days before the job starts so that workers and plant are not sitting around waiting.

The work being done should mirror the programme of work. The following shows how the programme is run over the 20-day contract period:

Day 1

The first thing to do on Day 1 is to set the site up and start digging the foundations. On all sites there are inductions and sometimes site meetings to discuss the contract. Calls should be made to all suppliers and sub-contractors to ensure that they are available to deliver/work on the required days, which will avoid delays to the programme through no-shows of materials or labour.

Day 2

Digging of the foundations should continue and be completed. The delivery of the materials and plant required for the foundations should arrive.

Day 3

The labour required to lay the foundations should arrive first thing in the morning and the laying of the foundations should be started. The materials and plant required for the cabling, piping, etc. should arrive.

Day 4

The site starts to get busier and the foundations should be finished today. The cabling and piping of the services should start.

Day 5

The cabling should be finished and ready to connect to the new structure. The materials for building the outer walls should arrive today, ready for Day 6.

Day 6

The bricklayers arrive on site and start to build the walls up to DPC height.

Day 7

The bricklayers finish building up to DPC height.

Day 8

The outer walls start to be built to finished height and the materials for laying the concrete floor should arrive.

Day 9

The outer walls continue to be built up and the concrete ground floor can be started.

Day 10

The outer walls continue to be built up and the concrete ground floor will be finished.

The programme should ideally be monitored every day. Now you are at the halfway stage, it is a good time to take stock: if the project is running behind, put your contingency plans into action now and if the project is ahead of schedule, see if materials, labour, etc. can be brought forward.

Days 10 and 11

The construction of the outer walls continues.

Day 12

The construction of the outer walls is completed and the materials for the roofing should arrive on site.

Day 13

The carpenters arrive on site and fit the roof trusses.

Day 14

The roofing contractors arrive on site and start to clad the roof with felt, tiles, etc. The materials for laying the floor should arrive.

Day 15

The site gets busy again as the roofers finish cladding the roof and the carpenters are back on site to lay the upper floor joists. The external doors and windows should arrive on site along with the materials for plumbing and electrical first fix.

Day 16

The flooring will be laid and the exterior windows and doors will start to be fitted. The plumber and electrician will arrive on site and start first fitting the pipes, cables, etc. The staircase will arrive on site along with the garage door and materials for painting the exterior woodwork.

Day 17

The exterior windows and doors will be finished and the main garage door will be fitted. The staircase along with all handrails, etc. will also be fitted today and the plumber and electrician will have finished their first fix. The painters will start painting the exterior of the garage and materials for the stud partitions and second fix plumbing/electrical work will arrive on site.

Day 18

The stud partitions will be built and the walls and ceilings will all be clad with plasterboard. The painter will finish the exterior work and the plumber and electrician can start their second fix work. Materials for all the internal paint and woodwork will arrive on site.

Day 19

The cladding of all the walls, ceilings etc. will be finished, as will the second fix plumbing and electrical work. The internal doors and second fix carpentry will also be done today. The internal painting work will be started.

Day 20

The internal painting work will be finished and the site will be tidied up ready for the handover to the client.

Completing the job

Now that the contract has been completed and the site cleaned up, it is time to invoice the client for the work carried out (on some larger jobs you may decide to use stage payments). For a company just starting out, this is a good time to take pictures of the work (asking the client's permission first), to show future clients the standard achieved. This is also the time to file all the drawings, invoices, costs, etc. for future reference.

On the job: Project managing

Bill and Nelson are running the job of building the garage. They look at the programme of work and notice that the painters aren't required until day 17. Bill thinks it might be a good idea to get the painters on site a couple of days early so that they can be painting as soon as possible. Nelson disagrees and thinks that could cause problems. Who do you agree with? What could be the outcome of having the painters in early?

FAQ

What can be done if a client changes their mind and wants something different?

A variation order and confirmation notice should be used to track the changes. The contractor must be sure to record all extra or different work, which will not have been included in the original pricing.

How many different plans/drawings do I need?

As many as are required to show the contractor what is to be built and how.

What does 500 people days mean?

500 people days refer to the number of people on site for a certain number of days. For example if you have 500 people on site for one day, that is 500 people days; if you have 1 person on site for 500 days then that is also 500 people days.

FAQ

What if it rains for the entire 20-day duration of the job?

The job would be seriously behind schedule. You can't plan for the weather in this country, but it would be unwise to start this job during a rainy season. There are companies that can provide scaffolding with a fitted canopy to protect the work area, which would be ideal for a job of this size. Larger jobs have longer programmes, and when they are drawn up they are made more flexible to allow for a lot of rainy days.

Knowledge check

1. What is the local authority responsible for?

2. Give a brief description of either of the two methods that can be used by a client when planning a contract.

3. What else can be offered with a tender to help win a contract?

4. State three pieces of information that can be obtained from a programme of work.

5. Why is it a good idea to take photos of the finished job?

chapter 5

Advanced first fixing

OVERVIEW

Carpenters and joiners will carry out many different types of work during their career, and this chapter looks at the first fix aspect of that work. First fixing is work that is usually done before the plastering, including:

- windows
- stairs
- joist and stud coverings
- temporary work and formwork.

All of the above are covered comprehensively in *Carpentry and Joinery NVQ and Technical Certificate Level 2* Chapter 9. In this book we will look more closely at two of these areas of work, as well as introducing two new topics.

The areas we will be looking at more closely are:

- windows
- stairs.

The two new topics this chapter will cover are:

- joist and stud coverings (plasterboard)
- temporary work and formwork.

Windows

Windows are designed with three main factors in mind:

- allowing natural light into the room/dwelling
- protecting the room/dwelling from weather
- keeping the heat in and cold out of a room/dwelling.

With windows it is also important to take both planning and conservation into consideration. Planning is vital when installing new windows as in some cases the type and style of window being fitted must be the same as the ones being replaced – this is especially important when working on listed buildings.

Window parts

Different types of windows have different parts, but the majority will have:

- **sill** – the bottom horizontal member of the frame
- **head** – the top horizontal member of the frame
- **jambs** – the outside vertical members of the frame
- **mullion** – intermediate vertical member between the head and sill
- **transom** – intermediate horizontal member between the jambs.

If a window has a sash, the sash parts will consist of:

- **top rail** – the top horizontal part of the sash frame
- **bottom rail** – the bottom horizontal part of the sash frame
- **stiles** – the outer vertical parts of the sash frame
- **glazing bars** – intermediate horizontal and vertical members of the sash frame.

There are various different styles of window, the most common types being:

- **fixed glazing** – a frame with no opening sashes with the glass fixed directly into it
- **box frame or sliding sash window** – a window with sliding sashes
- **casement window** – a window with sashes that are hung either on the top or side and can be traditional or storm proof
- **bay window** – window that projects out from the building
- **bow window** – similar to bay windows but more segmental and rounded.

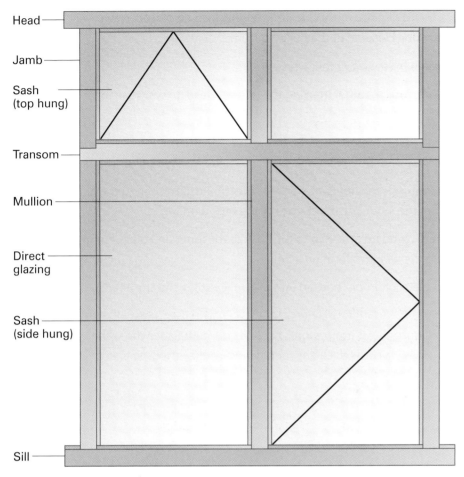

Head

Jamb

Sash
(top hung)

Transom

Mullion

Direct
glazing

Sash
(side hung)

Sill

Figure 5.1 Casement window

Casement windows are covered in *Carpentry and Joinery NVQ and Technical Certificate Level ²* and fixed glazing is simply a frame with a piece of glass fitted into it, so here we will concentrate on the box frame, bay and bow windows.

Box frame sliding sash window

Often referred to as a box sash window, this is rarely used nowadays – the casement window is preferred as it is easier to manufacture and maintain. Box sash windows are mainly fitted in listed buildings or in buildings where like-for-like replacements are to be used. Sometimes box sash windows are fitted because the client prefers them.

The box sash window is constructed with a frame and two sashes, with the top sash sliding down and the bottom sash sliding up. Traditionally the sashes work via pulleys, with lead weights attached to act as a counterbalance. The weights must be correctly balanced so that the sashes will slide with the minimum effort and stay in the required position: for the top sash, they must be only slightly heavier than the weight of the glazed sash; for the bottom sash, slightly lighter than the weight of the glazed sash.

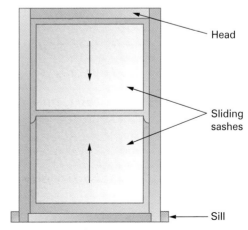

Head

Sliding
sashes

Sill

Figure 5.2 Box frame window

Most pulley-controlled systems now use cast-iron weights instead of lead. Newer box sash windows use helical springs instead of the pulley system.

The parts of a box sash window are as follows:

- **head** – the top horizontal member housing the parting bead, to which an inner and outer lining are fixed

- **pulley stiles** – the vertical members of the frame housing the parting bead, to which inner and outer linings are fixed. The pulley stiles act as a guide for the sliding sashes

- **inner and outer linings** – the horizontal and vertical boards attached to the head and pulley stiles to form the inner and outer sides of the boxed frame

- **back lining** – a vertical board attached to the back of the inner and outer linings on the side and enclosing the weights, forming a box

- **parting bead** – a piece of timber housed into the pulley stiles and the head, separating the sliding sashes

- **sash weights** – cylindrical cast iron weights used to counterbalance the sashes. These are attached to the sashes via cords or chains running over pulleys housed into the top of the pulley stiles

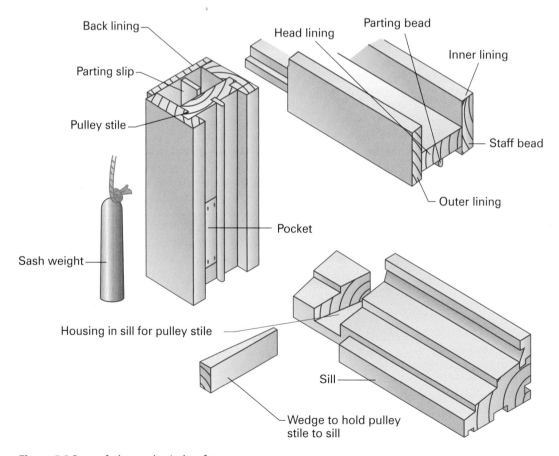

Figure 5.3 Parts of a box sash window frame

- **sill** – the bottom horizontal part of the frame housing the pulley stiles and vertical inner and outer linings

- **staff bead** – a piece of timber fixed to the inner lining, keeping the lower sash in place

- **parting slip** – a piece of timber fixed into the back of the pulley stiles, used to keep the sash weights apart

- **pockets** – openings in the pulley stile that give access to the sash cord and weights, allowing the window to be fitted and maintained.

The sash in a box frame is made slightly differently to normal sashes, and has the following parts:

- **bottom rail** – the bottom horizontal part of the bottom sash

- **meeting rails** – the top horizontal part of the bottom sash and the bottom horizontal part of the top sash

- **top rail** – the top horizontal part of the top sash

- **stiles** – the outer vertical members of the sashes.

Installing a box sash window

This can be done in one go with the sashes and weights already fixed, but for the purposes of this book we will show a more traditional method.

1. First strip the frame down and have the sashes, parting bead, staff bead and pockets removed.

2. Then fix the frame into the opening, ensuring that it is both plumb and level, and fit the sash cord and sashes (the fitting of the sash cord to the sashes will be covered in Chapter 8 page 179).

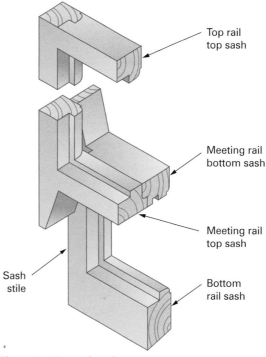

Figure 5.4 Box sash sashes

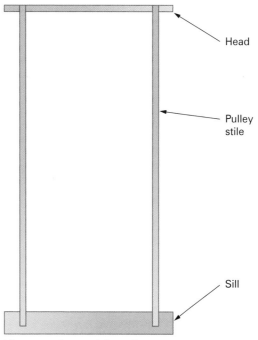

Figure 5.5 Stripped down frame

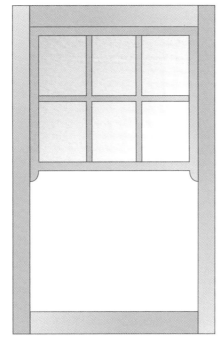

Figure 5.6 Top sash in place

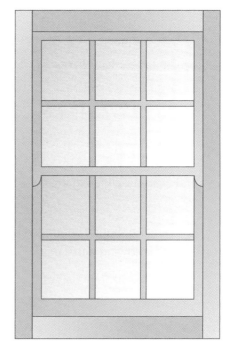

Figure 5.7 Top and bottom sashes in place

3. Next slide the top sash into place and fit the parting bead, to keep the top sash in place.

4. Now fit the bottom sash, then the staff bead, to keep the bottom sash in place.

5. Finally, seal the outside with a suitable sealer, fit the ironmongery and make good to the inside of the window.

The marking out of windows will be covered in Chapter 10 and the maintenance of a box sash window in Chapter 8.

Bay window

Bay windows project out from the face of the main outer walls of the dwelling. There are several different styles, but only three ways of building or incorporating one into a house.

Method 1 Pre-built bay

This is the simplest method, as it is part of the original design of the house and is constructed when the base walls of the house are. The area in the bay contributes to the size of the interior of the room.

Method 2 Built bay

With this method a bay is added onto the original structure of the house. The builder may open up the existing wall and build the bay into it, giving a larger room, or leave the wall and have a large window board area that can be used for storage – the choice depends on both the bay size and the client's wishes.

Remember

When replacing a bay, it is important to ensure that the roof is adequately supported *before* removing the old window

Method 3 Supported bay

This is only used for small bays with a small **projection**. Here a support system carries the weight of the bay, but no wall is built. The room is no bigger, but there will be a larger window board area.

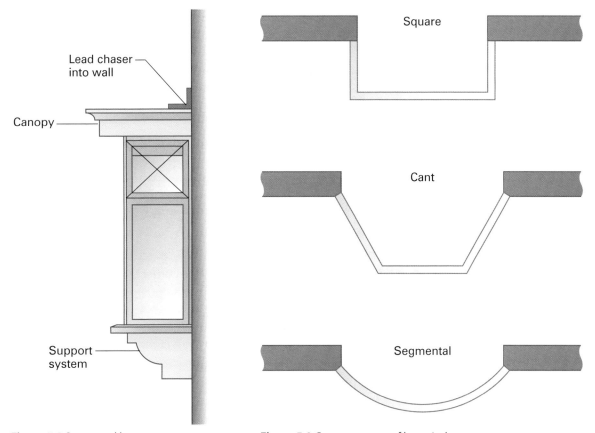

Figure 5.8 Supported bay

Figure 5.9 Common types of bay window

Types of bay window

As you will know from Level 2, several types of bay window can be constructed. The choice of style will depend on the size and shape of the opening or existing bay.

Bay windows are usually made up as a series of windows joined together rather than a single window. Bay windows are joined together by either manufacturing the windows to allow them to be screwed together or by using pre-made posts.

If the window is a square bay, the windows can be joined together like this:

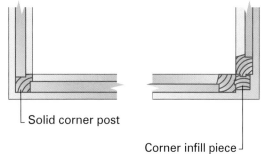

Figure 5.10 Square bay joints

If the window is a splayed or cant bay, they can be joined together as shown in Figure 5.11.

There may need to be a joint along the front of a bay window (usually only on very large windows), in which case the windows are joined together as shown in Figure 5.12.

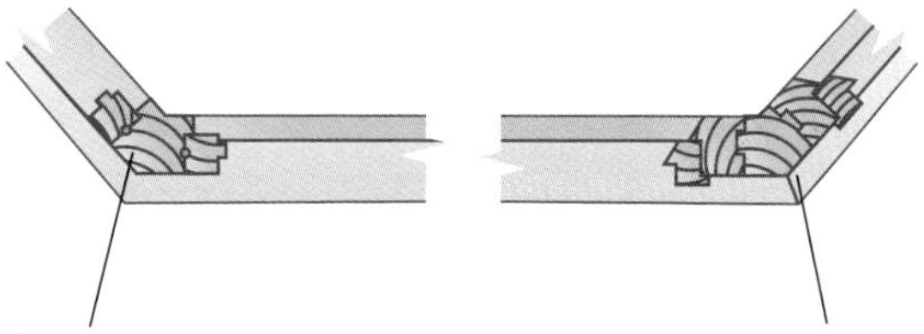

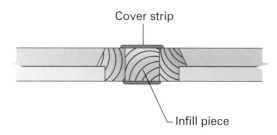

Figure 5.11 Cant bay joints

Figure 5.12 Lengthening bay joints

Remember

When removing an old window, the less damage you do, the less you will have to make good later

Remember

When fitting the first window, it is a good idea to run a small bead of **silicone** between the windows and corner posts to help with waterproofing

Installing a bay window

The first thing to do is to remove the old window. The old window will be load bearing, so the roof above it must be suitably supported before it is removed.

Next, fit and level the windowsill. Then fit one of the end windows, level and plumb it, then fix it to the wall.

Now fit the corner block and the front window, screwing it to the corner block and ensuring it is level, plumb and square to the first window. Do not fix the window to the sill or roof/wall above yet as it may need to be moved slightly to allow the last window to be fitted.

Next fit the final corner block and last window, screwing it to the wall, ensuring it is level, plumb and square to the front window. Now the windows can be fixed to the sill and roof/wall above and, if required, the cover strips at the joints can be fitted. Running a small bead of silicone along the back of the cover strip prior to fixing will help with waterproofing.

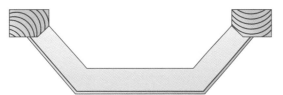

Figure 5.13 Fit the windowsill

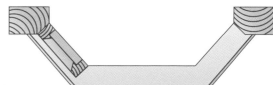

Figure 5.14 Fit the first window

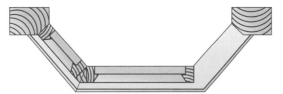

Figure 5.15 Fit the front window

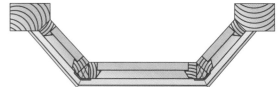

Figure 5.16 Window fitted including cover strips

To finish, glaze the window, fit any ironmongery, seal the exterior with silicone, fit the window board and make good to the inside.

Bow window

Bow windows are similar to **segmental** bay windows. They are made as a single window if the radius of the arch is not too great; otherwise they are made as separate windows.

The segmental bow window is fitted in the same way as the bay window, with the sill fitted first and the windows fitted from one side to the other. A bow window made as a single window is fitted in the same way as a normal window, with extra care being taken when handling the large window.

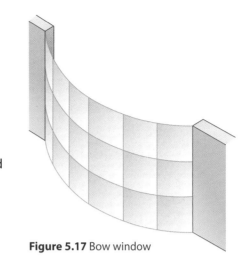

Figure 5.17 Bow window

Stairs

There are various different types of stairs, including:

- **straight flight** – where the stairs go straight from one floor to another (used where there is plenty of space)

- **quarter turn** – where the staircase turns 90 degrees while rising. This consists of two straight flights meeting at a platform or landing (usually in the middle). This type of stair is used where there is not much space or where the top landing needs to be in a different direction to the bottom flight of stairs

- **dog leg** or **half-turn stair** – where the staircase turns 180 degrees while rising. This stair can be constructed in two ways, the choice depending on the space available or the number of landings required

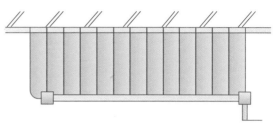

Figure 5.18 Straight flight

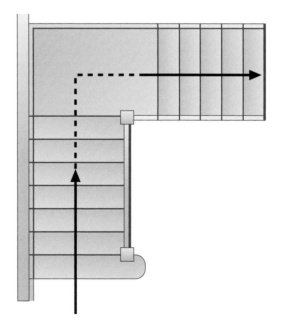

Figure 5.19 Quarter turn

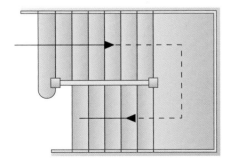

Figure 5.20 Dog leg

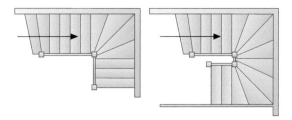

Figure 5.21 Winders

- **quarter/half turn with winders** – continually rising with **tapered** steps fitted at the turn but no platforms. This is used when there is a need for a direction change but no space for a landing

- **spiral staircase** – completely circular, turning continuously while rising. This is used when there is very little room.

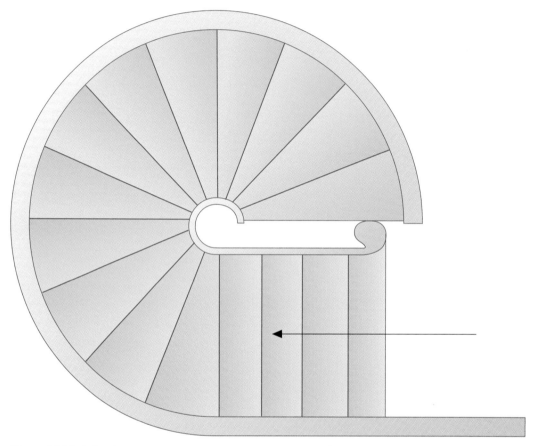

Figure 5.22 Spiral staircase

The straight flight is covered fully in *Carpentry and Joinery NVQ and Technical Certificate Level 2* page 292, while spiral staircases are usually metallic and fitted by specialists, so here we will look at the set-up of a turn in stairs and the use of winders within a turn.

Before we look at these staircases in more detail it is important to remind ourselves of the regulations that govern staircases, which will be familiar to you from Level 2.

Description of stair (to meet)	Max rise (mm)	Min going (mm)	Range (to meet pitch limitation) (mm)
Private stair	220	220	155–220 rise with 245–260 going or 165–200 rise with 220–350 going
Common stair	190	350	155–190 rise with 240–320 going
Stairway in institutional building (except stairs used by staff)	180	280	
Stairway in assembly area (except areas under 100m^2)	180	250	
Any other stairway	190	250	

Table 5.1 Regulations for rise and going

Description of stair	Minimum balustrade height	
	Flight	Landing
Private stair	840	900
Common stairway	900	1000
Other stairway	900	1100

Table 5.2 Regulations for minimum heights

Description of stair	Minimum width (mm)
Private stair giving access to one room only (except kitchen and living room)	600
Other private stair	800
Common stair	900
Stairway in institutional building (except stairs used only by staff)	1000
Stairway in assembly area (except areas under 100m square)	1000
Other stairway serving an area that can be used by more than 50 people)	1000
Any other stairway	800

Table 5.3 Regulations for width of stairs

For a list of components and their definitions, see *Carpentry and Joinery NVQ and Technical Certificate Level 2,* pages 294–295.

Quarter/half-turn staircase
Measuring out for a quarter-turn staircase

Look at this as two separate staircases. The first runs from the floor to the landing, and the second from the landing to the destination (top landing). When measuring out, it is important to get the landing sizes correct for the staircase to fit.

There are two methods of measuring out for and fitting a quarter-turn stair.

Method 1

The landing is built first, then each staircase is measured separately. The measurements are then sent to a joiner's shop and the staircases are made. The made staircases are delivered and fitted, with the handrails and **balustrade** fitted at the same time. This method is often preferred as there is less chance of making a mistake.

Method 2

Here all the measurements are taken first and sent to the joiner's shop, so that the stairs can be made while the landing is built. Once delivered, the stairs can be fitted along with the handrails and balustrades. If the measurements are not accurate, this method can lead to mistakes.

Fitting a quarter-turn staircase

For this example, we will use the first method.

First build the landing to the correct dimensions and in line with Building Regulations. The landing should be built out of minimum 4″ × 2″ softwood timber, and ideally be constructed as shown.

Figure 5.23 Landing

Next fit the staircases (bottom staircase first).

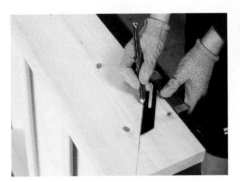

Step 1 Fix the wall string, cutting it off at floor level to suit the skirting height.

Step 2 Use a handsaw to cut the plumb cut on the string at the foot of the stairs.

Step 3 The staircase is now level to the floor.

Step 4 Steps 1 and 2 can be repeated at the top with the underside cut out to sit on to the floor trimmer and the top tread is cut away so that it sits on the trimmer.

Step 5 Fit the second staircase in the same way except that the cut at the bottom of the landing should be notched to sit on the floor.

Finally fit the newel posts, handrails, etc.

Step 6 Mortise the outer string into the newel posts at each end.

Step 7 Fix the newel posts in place. The bottom newel can be held in position using various methods depending on the composition of the floor. For rigidity and strength, the top newel should be notched over the trimmer joist and screwed or bolted to it.

Step 8 Fix the wall string to the wall in approximately four places below the steps, usually with 75mm screws and plugs. The balustrades and handrail can be fitted once the stairs are secure.

Step 9 Once the staircase has been fitted, it should be protected to prevent damage. Strips of hardboard should be pinned to the top of each tread with a lath to ensure the nosing is protected. Use the same method to protect the newel posts.

A half-turn staircase should be built the same way, preferably with the landing being built first.

The use of winders in a turn

'Winders' are the tapered treads on a staircase that turns as it rises without using a landing. The first thing to note is that there are extra regulations regarding winders.

- The rise of the tapered steps must be the same as the rise of the other steps.

- The tapered step must not be less than 50mm at its narrowest point.

- The going of the tapered steps (measured at the centre of the steps) must be the same as the going of the other steps.

Setting out a staircase containing winders

The number of winders in a tapered stair with a quarter or half turn depends on the size of the staircase. A staircase that is wide with a large going will have more winders than a small staircase. Our example will use a three-winder staircase.

First do a scale plan drawing of the stairs, to help you determine the size and going of the winders.

Next comes the construction. This is fundamentally the same as for a straight flight (covered in *Carpentry and Joinery NVQ and Technical Certificate Level* [2]), but with a few differences:

- the **strings** are shaped differently, as the wall strings need to be larger to allow for the rise during the change of direction. Figure 5.25 gives a typical example of how the wall strings will look

- the **newel post** has to be housed out accordingly as well, like this:

Now fit the stairs. The made staircase should be delivered to site in two separate parts, with the newel post and winders taken out (to allow for ease of transportation and prevent damage). The simplest way to fit it is to treat it as two separate quarter-turn stairs.

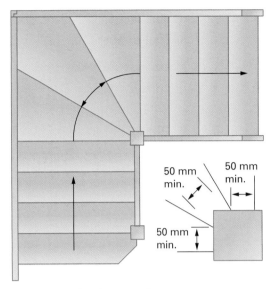

50 mm min.
50 mm min.
50 mm min.

Figure 5.24 Plan drawing of stairs

Figure 5.25 Wall strings for winders

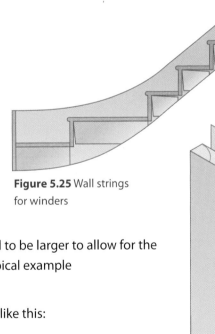

Figure 5.26 Newel post development to allow for winders

Fix the top section in place, ensuring that the cuts at the top are correct and the treads are level. Fit the newel post to support the open end of the stair. Position the winders and slide the bottom section into place. Check and adjust for level, then fix the bottom section, handrails and balustrades.

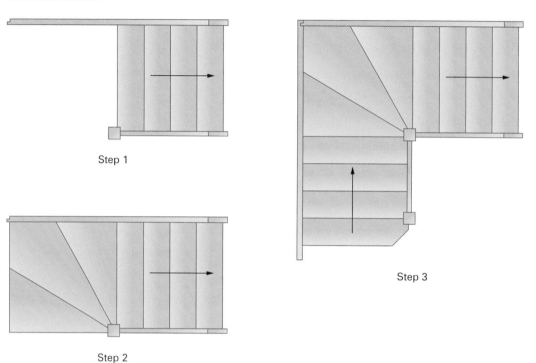

Step 1

Step 3

Step 2

Figure 5.27 Three stages of winders being fitted

Joist and stud coverings

Joists are decked with a suitable flooring material, and the underside of the joists is usually covered with plasterboard. Stud partitions are also usually covered with plasterboard, with the occasional exception. For example, areas such as shower rooms are likely to be tiled, so a sturdier board such as WBP plywood is used.

Joist and stud coverings have come a long way in the past few years, and the traditional method of lath and plaster is now only really used in listed buildings.

Lath and plaster

There are two main ways of plasterboarding a room:

Method 1

The plasterboard is fixed to the stud or joist with the back face of the plasterboard showing. The plasterer covers the whole wall or ceiling with a thin skim of plaster, leaving a smooth finish.

Plastered wall

Taped wall

Method 2

The plasterboard is fixed with the front face showing, and the plasterer uses a special tape to cover any joints, then a ready mixed filler is applied over the tape and is used to fill in the nail and screw holes. Once the taped area is dry, the plasterer then gives the area a light sand to even it out.

Plasterboard comes with a choice of two different edges, and the right edge must be used:

- square edge – for use with Method 1, the whole wall/ceiling is plastered

- tapered edge – for use with Method 2, so that the plasterer can fix the jointing tape.

As well having the correct edging, the right type of plasterboard must be used.

Various types of plasterboard are available so a colour coding system is used (NB this may vary from manufacturer to manufacturer):

- fire resistant plasterboard (usually red or pink) – used to give fire resistance between rooms such as party walls

- moisture resistant plasterboard (usually green) – used in areas where it will be subjected to moisture such as bathrooms and kitchens

- vapour resistant plasterboard – with a thin layer of metal foil attached to the back face to act as a vapour barrier, usually used on outside walls or the underside of a flat roof

- sound resistant plasterboard – used to reduce the transfer of noise between rooms and can also help acoustics, used in places such as cinemas

- thermal check board – comes with a foam or polystyrene backing, so used to prevent heat loss and give good thermal insulation

- tough plasterboard – stronger than average plasterboard, it will stand up to more impact, used in public places such as schools and hospitals.

Sometimes plasterboard is doubled up – the walls are 'double sheeted' – to give fire resistance and sound insulation without using special plasterboard.

As well as a variety of types plasterboard comes in a variety of sizes, though the most standard size is 2400 × 1200 × 12.5mm.

Cutting plasterboard

Plasterboard can be cut in two main ways: with a plasterboard saw or with a craft knife. The plasterboard saw can be used when accuracy is needed but most jobs will be touched up with a bit of plaster, so accuracy is not vital. Using a craft knife is certainly quicker.

This is how to cut plasterboard with a craft knife:

Step 1 Mark the line to be cut with a pencil.

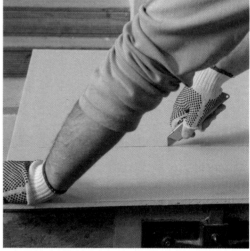

Step 2 Score evenly along the line with the craft knife ensuring hands, legs, etc. are out of the way of the blade.

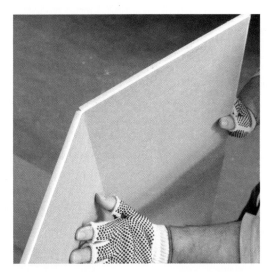

Step 3 Turn the board round and give the area behind the cut a slap, to split the board in two.

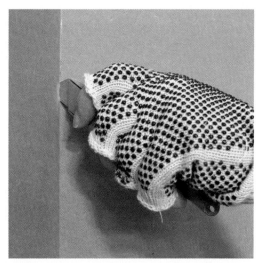

Step 4 Run the knife along the back of the cut to separate the two pieces of board.

Step 5 Trim the cut with a slight back bevel to give a neater finish.

Fixing plasterboard

Plasterboard can be fixed by three different methods, which are:

- **nail**

Plasterboard can be fixed using **clout** or special plasterboard nails. When nailing, make sure the nail is driven below the surface with the hammer, leaving a dimple or hollow that can be filled by plaster later. The nails used must be galvanised: if not, when the plaster is put on to fill the hollows, the nails will rust and will 'bleed' – the rust from the nails will stain the plaster and can even show through several coats of paint.

- **screw**

Screws can be used but these too must be galvanised to prevent bleeding.

- **dot and dab**

This is used where the plasterboard is fitted to a pre-plastered or flat surface such as block work. With dot and dab, there is no need for a stud or frame to fix the plasterboard, thus increasing the room size (it is often used in stairwells where space is limited). Dot and dab involves mixing up plaster and dabbing it onto the back of the board, then pushing the board directly onto the wall.

Temporary work and formwork

As the name suggests, temporary work is work that is only there for a short time and will be removed prior to the finish. There are various types of temporary work and formwork such as:

- **support work** – can take several shapes such as arch centres, which are used to support brick arches

- **shuttering** – where a frame is built, into which concrete is poured. Shuttering is used on jobs like concrete columns and pillars

- **shoring** – where an existing building or component is supported to prevent falling during modification work. For example, where a ceiling or roof is supported, when a lintel or load-bearing partition is being moved or modified, etc.

- **mould boxes** – a timber box is made to a specific shape, into which concrete is poured. Once the concrete has set the box is removed, leaving a pre-cast member such as a windowsill or even a staircase.

We will now look at an example of each of these.

Support work

This is where something is made to support the work while it progresses, a perfect example being an arch centre. An arch centre is used to support the brickwork while the arch is being built, and is usually left *in situ* until the keystone is fitted and the cement has set.

Arch centres are usually made from timber. If the span of the opening is larger than 1.5m, a stronger metal centre is used. The arch centre has several parts:

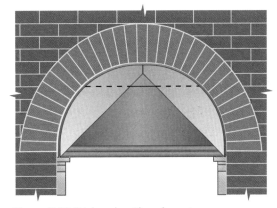

Figure 5.28 Brick arch with arch centre

- **ribs** – usually made from 19mm plywood, MDF or 22mm solid timber, these form the outer part of the centre, giving the outline for the rest of the framework

- **ties** – horizontal members attached at the bottom of the ribs to prevent the load pushing the ribs out of shape

- **struts** – vertical members used to transfer the weight and spread the load

- **bearers** – flat members fixed to the bottom of the ribs to tie the ribs together

- **bracing** – timber fixed at an angle on the inside of the arch to give support and keep the arch square

- **laggings** – either thin sheets of hardboard or plywood nailed across the top of the ribs, or small timber laths nailed at right angles to the ribs.

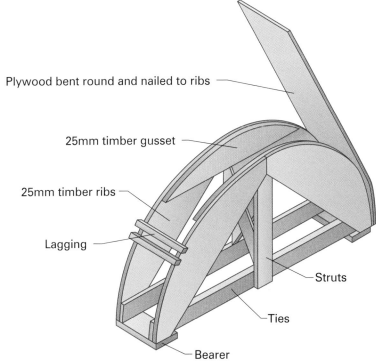

Plywood bent round and nailed to ribs

25mm timber gusset

25mm timber ribs

Lagging

Struts

Ties

Bearer

Figure 5.29 Arch parts

When setting out for an arch centre you need to know the span, arch radius and so on. You can get this information from the bricklayer or from drawings.

First draw out the arch full size. Use the drawing to make templates out of hardboard, which can then be traced onto the ribs. The way the ribs are made up depends on the size and radius of the arch: some arches use two ribs per side while others use four.

Now nail the ribs together using gusset plates, fixing them where the gusset overlaps the ridge, then fix the struts and ties. Usually both ribs are made as a pair and checked together against the drawing made earlier.

The bracing is fixed next, bringing both ribs together, then the bearers can be fitted.

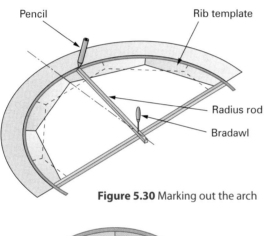

Pencil · Rib template · Radius rod · Bradawl

Figure 5.30 Marking out the arch

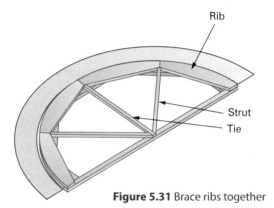

Rib · Strut · Tie

Figure 5.31 Brace ribs together

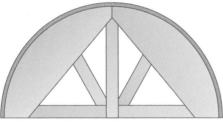

2-rib set up

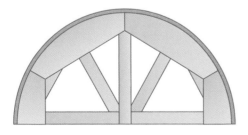

4-rib set up

Figure 5.32 Two-rib and four-rib set-ups

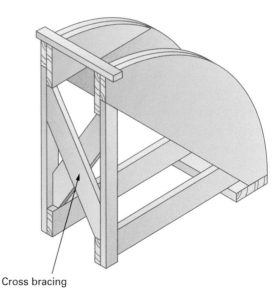

Cross bracing

Figure 5.33 Fit bracing and bearer

Finally fit the lagging, which is nailed down through the tops of the ribs.

Now the arch centre is ready to be used.

Shuttering

Shuttering is used mostly on columns and pillars. The method for shuttering a column is as follows:

First form an **upstand** (or **kicker**) for the column. This helps locate the position of the column and reduces the amount of concrete lost at the base of the column when it is poured.

The kickers are formed using a frame fixed to the floor, which is filled with concrete, and a **ballistic tool**. When casting the kickers it is vital that they are in the correct position.

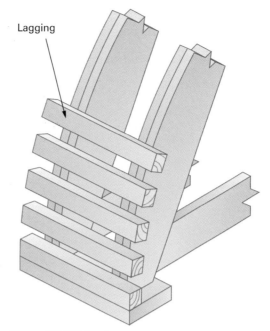

Lagging

Figure 5.34 Fit lagging

> **Definition**
>
> **Upstand** or **kicker** – a concrete cast base, which is used to help locate a pillar of column in the correct place

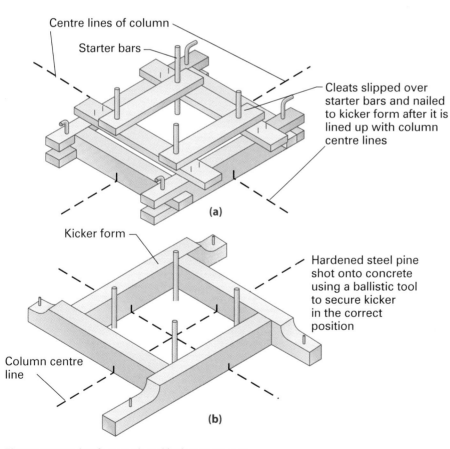

Centre lines of column

Starter bars

Cleats slipped over starter bars and nailed to kicker form after it is lined up with column centre lines

(a)

Kicker form

Hardened steel pine shot onto concrete using a ballistic tool to secure kicker in the correct position

Column centre line

(b)

Figure 5.35 Kicker formwork and kicker on its own

Next fit the shuttering round where the column will go. The shuttering is made up of a series of timber frames clad on one side with plywood, and must be fixed together with screws.

Now fit a series of clamps to the shutter to prevent the concrete bursting through the sides.

The whole structure must now be levelled, plumbed and adequately braced with diagonal bracing to prevent any movement.

Pour the concrete. Once it has set, remove the bracing, clamps and shuttering.

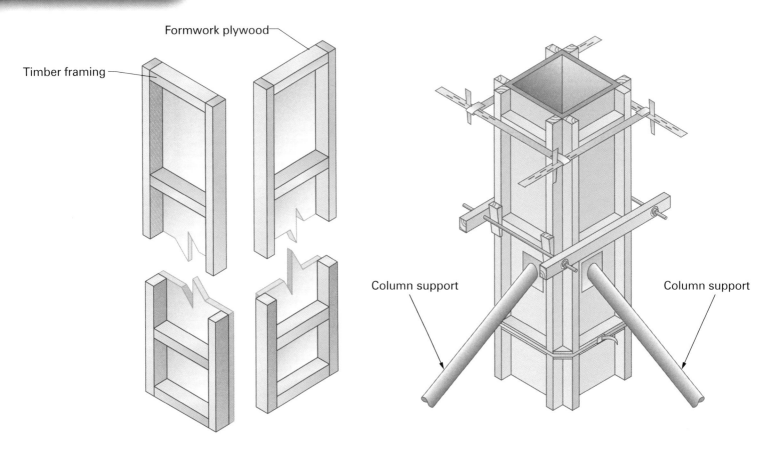

Formwork plywood

Timber framing

Column support

Column support

Figure 5.36 Shuttering

Figure 5.37 Braced column

Shoring

Shoring is used where an existing building or component is supported to prevent falling or collapse during modification or removal work. Our example here is the removal of a load-bearing stud partition.

Proper shoring is vital, and is most commonly done with **telescopic** props adjusted to take the weight of the floor above.

Removal of a load-bearing partition

First determine which way the joists on the upper floor run, to make sure your props will support the weight.

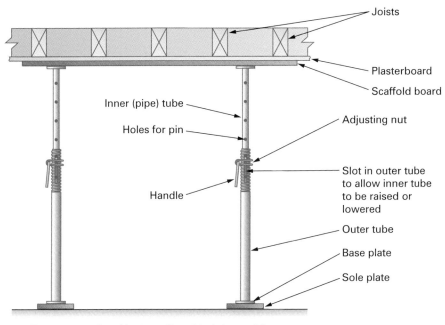

Correct way of positioning adjustable (telescopic) prop

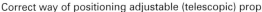

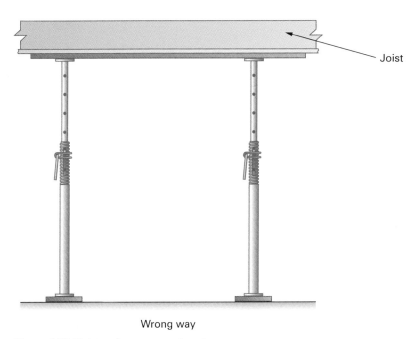

Wrong way

Figure 5.38 Right and wrong way to set up props

Ideally, the props should be placed as close to the partition as possible without being in the way of operations, with the upper floor propped on both sides of the partition. The props should be placed onto a board, acting as a sole plate, and a board should also be positioned where they meet the ceiling, to spread the load.

Now the work can commence. The props should only be removed once the work is complete and a suitable support system such as a steel beam is in place.

Mould boxes

Mould boxes are similar to shuttering, but are used to form items that are more complicated. Mould boxes can be dismantled, re-assembled and used again.

A simple mould box for a windowsill would look something like Figure 5.39.

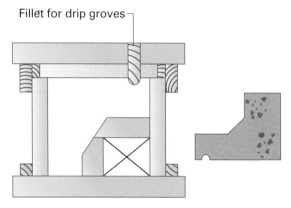

Fillet for drip groves

Figure 5.39 Windowsill mould box

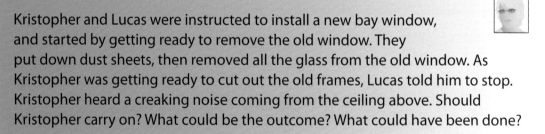

On the job: Removing a bay window

Kristopher and Lucas were instructed to install a new bay window, and started by getting ready to remove the old window. They put down dust sheets, then removed all the glass from the old window. As Kristopher was getting ready to cut out the old frames, Lucas told him to stop. Kristopher heard a creaking noise coming from the ceiling above. Should Kristopher carry on? What could be the outcome? What could have been done?

FAQ

I didn't know there were so many different types of plasterboard available. Which type should I use?

The type of plasterboard used should be stated by the client or in the architect's drawings. There are different types because there are so many different purposes that rooms are used for.

I have been told that I have nailed the plasterboard on the wrong way. Which way is the right way?

Plasterboard has two faces. The front face is usually white with no markings, while the back has the manufacturer's logo on it. The right way will depend on what is happening to the plasterboard. If it is only having the joints taped, then the white face must be shown; if the whole wall is being skimmed, the plasterers usually prefer to have the back face showing.

Knowledge check

1. State the three main factors to keep in mind when windows are designed.

2. Name four main parts of a window frame.

3. Name three main parts of a window sash.

4. Name three types/styles of window.

5. What type of weights are used in a box sash window?

6. How heavy should the weights balancing the top sash in a box sash window be?

7. Name three different types of bay window.

8. State two ways in which a turn in stairs can be achieved.

9. State the minimum rise and going for a private stair.

10. State the maximum rise and going for a common stair.

11. When using winders, what is the minimum width of the winder at its smallest point?

12. When covering studs, when would plywood be used rather than plasterboard?

13. What is the difference in use between square edge and tapered edge plasterboard?

14. List four different types of plasterboard and their uses.

15. What is the standard size of plasterboard?

16. What is dot and dab?

17. Why must galvanised nails/screws be used when fixing plasterboard?

18. Name three different types of temporary work.

chapter 6

Advanced second fixing

OVERVIEW

Second fixing covers all the carpentry work done after the plastering. This work has to be more precise than first fixing work as it is usually visible. Second fixing work involves:

- mouldings (skirting, architrave, etc.)
- door hanging
- ironmongery (screws, nails locks, etc.)
- encasing services (pipe boxes, bath panels, etc.)
- wall and floor units (kitchens, bathrooms, bedrooms, etc.).

All of the above are covered in *Carpentry and Joinery NVQ and Technical Certificate Level 2*, so the aim here is to expand your knowledge of some areas and to look at more complex aspects of the work. The topics we will be looking at are:

- external doors (including ironmongery)
- hanging an external door
- double doors (including ironmongery)
- sliding doors
- dado, picture rail and cornice
- wall panelling.

External doors

This section aims to provide the knowledge and understanding needed to select, hang and fix the required ironmongery to an external door.

Types of external door

External doors are always solid or framed – a flush or hollow door will not provide the required strength or security.

In some cases external doors are made from UPVC. UPVC doors require little maintenance and the locking system normally locks the door at three or four different locations, making it more secure. High quality external doors are usually made from hard-wearing hardwoods such as mahogany or oak, but can be made from softwoods such as pine.

External doors come in the same dimensions as internal doors except external doors are thicker (44mm rather than 40mm). On older properties external doors often have to be specially made as they tend not to be of standard size.

There are four main types of external door:

- framed, ledged and braced
- panelled doors
- half/full glazed doors
- stable doors.

Did you know?

A framed, ledged and braced door comes with the bracing unattached. The bracing can then be attached to suit the side on which the door is hung

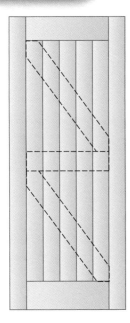
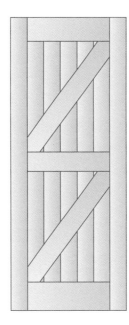

Front elevation   Back elevation

Figure 6.1 Front and back of a framed, ledged and braced door

Framed, ledged and braced doors

Framed, ledged and braced (FLB) doors consist of an outer frame clad on one side with tongued and grooved boarding, with a bracing on the back to support the door's weight.

FLB doors are usually used for gates and garages, and sometimes for back doors. When hanging an FLB door it is vital that the bracing is fitted in the correct way; if not, the door will start to sag and will not operate properly.

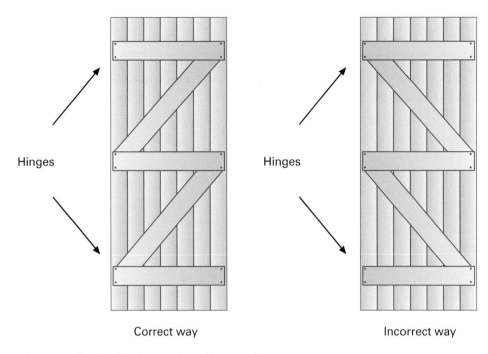

Figure 6.2 Bracing fitted correctly and incorrectly

Panelled doors

Panelled doors consist of a frame made up from stiles, rails, muntins and panels. Some panel doors are solid, but most front and back doors have a glazed section at the top to allow natural light into the room.

Half/full glazed doors

Half-glazed doors are panelled doors with the top half of the door glazed. These doors usually have diminished rails to give a larger glass area.

Full glazed doors come either fully glazed or with glazed top and bottom panels, separated by a middle rail.

Full glazed doors are used mainly for French doors or back doors where there is no need for a letterbox.

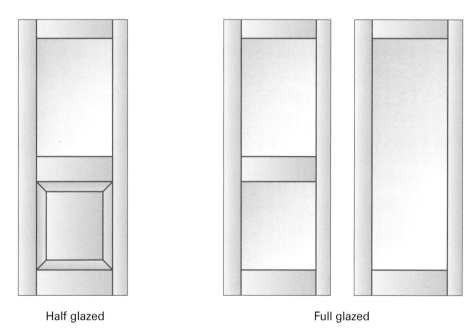

Figure 6.3 Half-glazed door and two types of full glazed door

Stable doors

As their name implies, stable doors are modelled on the doors for horses' stables and they are now most commonly used in country or farm properties. A stable door consists of two doors hung on the same frame, with the top part opening independently of the bottom. The make-up of a stable door is similar to the framed, ledged and braced door, but the middle rail will be split and rebated as shown in Figure 6.4.

To hang a stable door, first secure the two leaves together with temporary fixings, then hang just like any other door, remembering that four hinges are used instead of three.

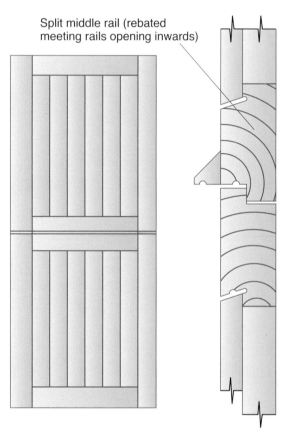

Split middle rail (rebated meeting rails opening inwards)

Figure 6.4 Stable door with section showing the rebate on the middle rails

Hanging an external door

External doors usually open inwards, into the building. Where a building opens directly onto a street, this prevents the door knocking into unsuspecting passers-by, but, even where there is a front garden, having a door that opens outwards is not good practice as callers will need to move out of the way to allow the door to be opened. Externally opening front doors are usually only used where there is limited space, or where there is another door nearby which affects the front door's usage.

Hanging an exterior door is largely the same as hanging an interior door (see *Carpentry and Joinery NVQ and Technical Certificate Level 2* page 338) except that the weight of the door requires three hinges sited into a frame rather than a lining. Because of this, there is usually a threshold or sill at the bottom of the doorframe.

If a water bar is fitted into the threshold to prevent water entering the dwelling, you will need to rebate the bottom of the door to allow it to open over the water bar. The way the door is then hung will depend on which side of the door is rebated. A weatherboard must also be fitted to the bottom of an external door, to stop driving rain entering the premises.

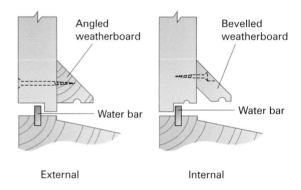

Figure 6.5 Internally and externally opening doors

As an exterior door is exposed to the elements it is important that the opening is draught proofed: a draught proofing strip can be fitted to the frame, or draught proofing can be fitted to the side of the door.

Ironmongery for an external door

An external door requires more ironmongery than an internal door, and may need:

- hinges
- letter plate
- mortise lock/latch
- mortise dead lock
- cylinder night latch
- spy hole
- security chains.

Hinges

External door hinges are usually butt hinges, though framed, ledged and braced doors often use T hinges. Three 4" butt hinges are usually sufficient, though it is advisable to use security hinges (hinges with a small steel rod fixed to one leaf, with a hole on the other leaf) to prevent the door being forced at the hinge side.

Letter plate

The position of a letter plate depends on the type of door. Letter plates are usually fitted into the middle rail, but could be fitted in the bottom rail (for full glazed doors) or even the stile (with the letter plate fitted vertically).

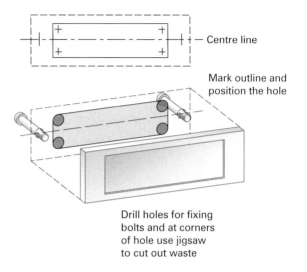

Centre line

Mark outline and position the hole

Drill holes for fixing bolts and at corners of hole use jigsaw to cut out waste

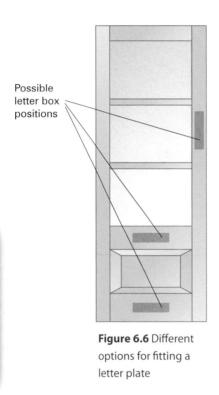

Possible letter box positions

Figure 6.6 Different options for fitting a letter plate

A letter plate can be fitted either by drilling a series of holes and cutting out the shape with a jigsaw, or by using a router with a guide.

Once the opening has been formed, the letter plate can be screwed into place.

Mortise lock/latch

Mortise locks are locks housed into the stile and are fitted as follows:

Step 1 Mark out the position for the lock (usually 900mm from the floor to the centre of the spindle) and mark the width and thickness of the lock on the stile.

Step 2 Using the correct sized auger/flat bit, drill a series of holes to the correct depth.

Step 3 Using a sharp chisel remove the excess timber, leaving a neat opening.

Step 4 Mark around the faceplate and with a sharp chisel, remove the timber so that the faceplate sits flush with the stile.

Step 5 Mark where the spindle and keyhole are, then drill out to allow them both to be fitted and operate properly.

Step 6 Fix the lock and handles in place.

Step 7 Mark the position of the striking plate on the doorframe, then house it into the frame.

Step 8 Check that the lock operates freely.

Basic security ironmongery

A mortise lock on its own does not usually provide sufficient security for an exterior door, so most doors will also have one of the following:

- **mortise deadlock** – fitted like a mortise lock except that it has no latch or handles, so an escutcheon is used to cover the keyhole opening. It is usually fitted three quarters of the way up the door.

- **cylinder night latch** – preferred to a mortise deadlock, as it does not weaken the door or frame as much. It is also usually fitted three quarters of the way up the door. The manufacturer will provide fitting instructions with the lock.

Spy hole

A spy hole is usually fitted to a door that is solid or has no glass, and is used as a security measure so that the occupier can see who is at the door without having to open the door.

To fit a spy hole, simply drill a hole of the correct size, unscrew the two pieces, place the outer part in the hole, then re-screw the inner part to the outer part.

Security chain

A security chain allows the door to be opened a little without allowing the person outside into the dwelling. The chain slides into a receiver, then as the door opens the chain tightens, stopping the door from opening too far.

Double doors

Here two doors are fitted within one single larger frame/lining, with 'meeting stiles': two stiles that meet in the middle, rebated so that one fits over the other.

Double doors are usually used where a number of people will be walking in the same direction, while double swing doors are used where people will be walking in different directions. Double doors allow traffic to flow without a 'bottleneck' effect, and let large items such as trolleys pass from one room to another. Double doors are used in places such as large offices, or public buildings such as schools and hospitals.

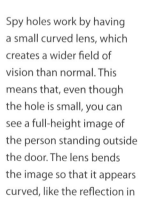

Did you know?

Spy holes work by having a small curved lens, which creates a wider field of vision than normal. This means that, even though the hole is small, you can see a full-height image of the person standing outside the door. The lens bends the image so that it appears curved, like the reflection in a Christmas bauble

Hanging double doors is the same as for any other door, though extra care must be taken to ensure that the stiles meet evenly, and that there is a suitable gap around the doors.

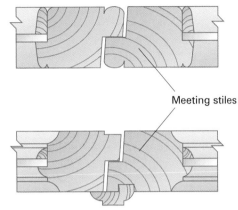

Meeting stiles

Figure 6.7 Meeting stiles

Extra ironmongery is required on double doors as follows:

Parliament hinge

- **parliament hinges** – these project from the face of the door and allow the door to open 180 degrees

- **rebated mortise lock** – similar to a standard mortise lock but the lock and striking plate are rebated to allow for the rebates in the meeting stiles

Rebated mortise lock

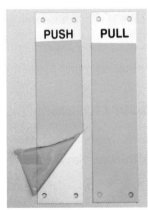

PUSH   PULL

Push or kick plates

- **push/kick plates** – fixed to the meeting stiles and the bottom rails to stop the doors getting damaged

- **door pull handles** – fixed to the meeting stiles on the opposite side from the push plates, these allow the door to be opened

- **barrel bolts** – usually fixed to one of the doors at the top or the bottom to secure the door when not in use.

Pull handles

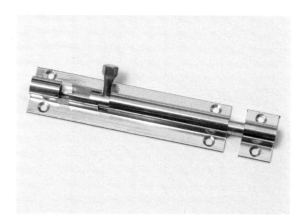

Barrel bolt

Most double doors also have door closers to ensure that they will close on their own, to prevent the spread of fire or draughts throughout the building.

There are four main types of door closer: overhead, concealed, floor springs and helical spring hinges. Floor springs and **helical hinges** will be looked at on page 108 when we deal with double swing doors, so for the moment we will look at concealed and overhead door closers.

Concealed door closers

Concealed door closers work through a spring and chain mechanism housed into a tube.

To fit a concealed door closer, you must house the tube into the edge of the hinge side of the door, with the tension-retaining plate fitted into the frame.

Overhead door closers

As the name implies, overhead door closers are fitted to either the top of the door or the frame above. They work through either a spring or a hydraulic system fitted inside a casing, with an arm to pull the door closed.

There are different strengths of overhead door closer to choose from depending on the size and weight of the door. Table 6.1 below shows the strengths available.

Power no.	Recommended door width	Door weight
1	750mm	20kg
2	850mm	40kg
3	950mm	60kg
4	1100mm	80kg
5	1250mm	100kg
6	1400mm	120kg
7	1600mm	160kg

Table 6.1 Strengths of overhead door closures

There are a number of different ways to fit an overhead door closer depending on where the door is situated. The two main ways are:

- fitting the main body of the closer to the door, with the arm attached to the frame

- inverting the closer so that the main body is fitted to the frame and the arm is fitted to the door.

Remember

When choosing an overhead door closer, you must take into account any air pressure from the wind. If the pressure is strong, you may require a more powerful closer

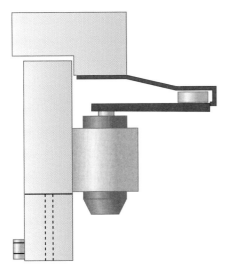
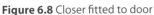

Figure 6.8 Closer fitted to door

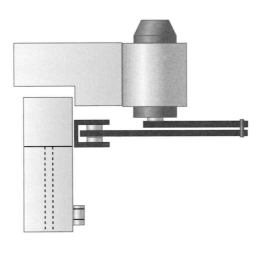

Figure 6.9 Closer fitted to frame

Door closers come with instructions and a template, to make fitting easier.

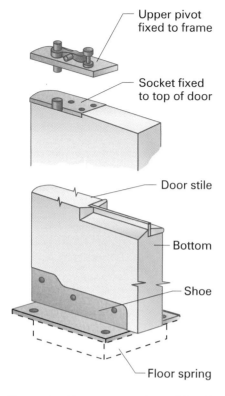

Figure 6.10 Floor spring and top of door fixing

Double swing doors need to open both ways to accommodate traffic going in both directions. The main difference work-wise between standard double doors and double swing doors is in the ironmongery. Double swing doors need special hinges, and must have some form of door closer fitted – and it is usually best to combine these, in one of two ways.

- **Helical spring hinges**

Helical spring hinges are three-leaf hinges with springs integrated into the barrels. The way the leaves are positioned allows the doors to open both ways. The hinges are fitted just like normal hinges and the tension on the springs are adjusted via a bar inserted into the hinge collar.

- **Floor springs**

Doors using floor springs are hung via the floor spring at the bottom and a pivot plate at the top. The floor spring is housed into the floor, and the bottom of the door is recessed to accept the shoe attached to the floor spring.

The pivot plate at the top is attached to the frame, and a socket is fixed to the top of the door.

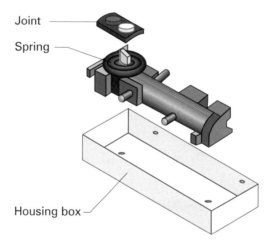

Figure 6.11 Helical floor springs

Sliding doors

Sliding doors are mainly used where there is no space for a door to open outwards. They are not hung in the traditional way with hinges, but use a track on which the door slides open along the face of the wall.

There are many different systems available that allow a door to be fitted in this way, but in most the basics are the same. The door is suspended from an overhead track, and slides on a series of rollers, while the bottom of the door is grooved so that it runs over a track or plastic guide. The overhead track section is usually encased in a pelmet, allowing the door to be detached easily for maintenance.

Remember

Installation differs from manufacturer to manufacturer, so it is best to follow the manufacturer's instructions

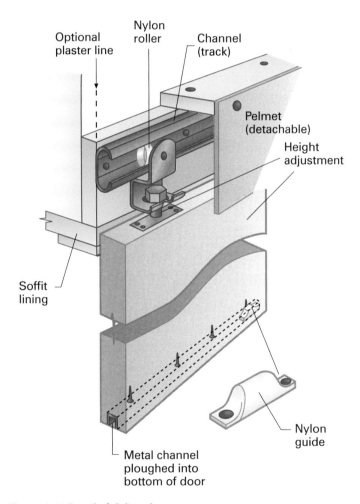

Figure 6.12 Detail of sliding door gear

Dado, picture rail and cornice

Carpentry and Joinery NVQ and Technical Certificate Level [2] covered architrave and skirting, so in this book we will expand our knowledge of mouldings to include dado rail, picture rail and cornice.

Mouldings in a room are positioned like this:

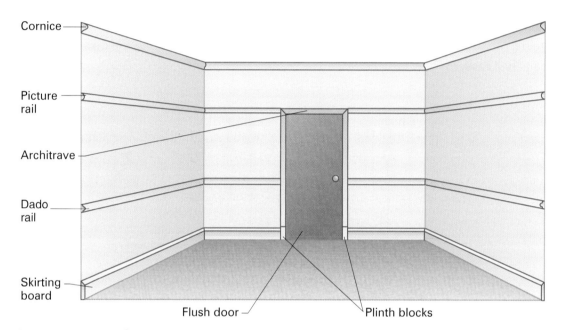

Figure 6.13 Positions of room mouldings

We will start at the top and work down, beginning with the cornice.

Cornice

Cornice is fitted where the top of the wall meets the ceiling. It is traditionally associated with plasterers, although some cornice today is made of timber and is fitted by the carpenter. Cornice can be fitted as a decoration piece or used to hide gaps or blemishes. As with most mouldings, various different designs are available.

Figure 6.14 Cornice profiles

Cornice is simple to install. You nail or screw it into the wall along the ceiling, taking care to ensure that it is running flat to both the wall and the ceiling. External joints are mitred, while internal joints can be either scribed or mitred.

Picture rail

Picture rail is usually fixed at the same height as the top architrave on a door, or just above or below this. Picture rail was used to hang paintings from, so the height used to be determined by the size of pictures being hung; nowadays picture rail is mainly for decoration. Picture rails also come in a variety of profiles.

Figure 6.15 Picture rail profiles

Picture rail is slightly more difficult to fix than skirting or cornice, as the rail must be fixed level. First mark a level line around the room to act as a guide for fitting the rail, then proceed as for any other moulding, with the joints either scribed or mitred.

Dado rail

Dado rail is fixed to the wall between the picture rail and the skirting, the exact height depending on how it is to be used. Dado rails were originally used to guard the walls from damage from chair backs, so would be set at the height of a chair. Today dado rails in domestic dwellings are mainly for decorative purposes, so the height is up to the owner's preference. Dado rails also come in a variety of profiles.

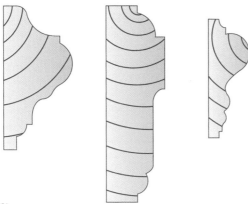

Figure 6.16 Dado rail profiles

You fix a dado rail in the same way as a picture rail, taking care to ensure that the rail is level and the joints used are scribes and mitres.

Did you know?

A form of dado rail is still used today in places such as hospitals to protect the walls from trolley damage

Wall panelling

Wall panelling provides a decorative finish to a room and can be found in places such as courts of law or executive offices. Wall panelling is usually set at one of three heights:

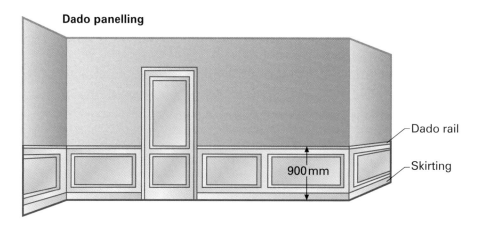

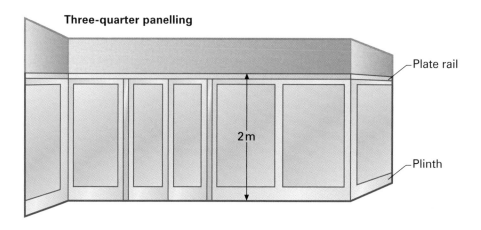

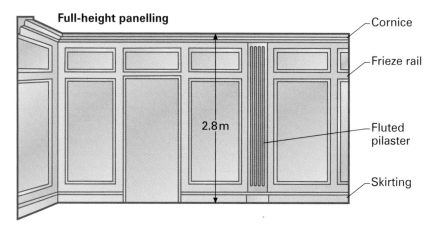

Figure 6.17 Three typical heights of panelling

- **dado panelling** – where the panelling runs to the height of the dado rail

- **three-quarter panelling** – where the panelling runs to the top of the door

- **full-height panelling** – where the panelling runs from floor to ceiling.

The panelling can be made up in a variety of ways, depending on the type and style of the house. The first thing to do is to make a frame or fix battens to the wall, onto which you can then fix your panelling. Once the panelling is in place, fix the capping pieces and skirting to finish the panelling off.

These three examples of dado panelling give you an idea of what is involved.

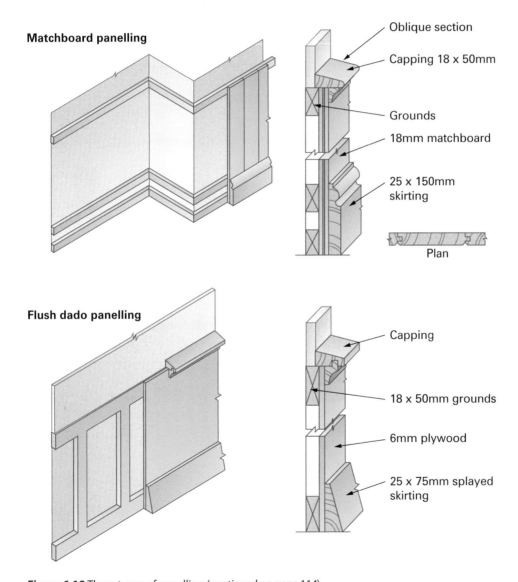

Matchboard panelling

Oblique section

Capping 18 x 50mm

Grounds

18mm matchboard

25 x 150mm skirting

Plan

Flush dado panelling

Capping

18 x 50mm grounds

6mm plywood

25 x 75mm splayed skirting

Figure 6.18 Three types of panelling (*continued on page 114*)

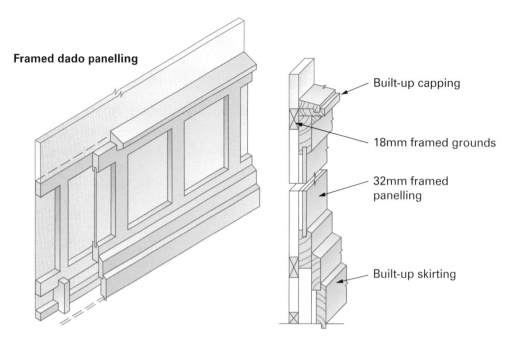

Framed dado panelling

Built-up capping

18mm framed grounds

32mm framed panelling

Built-up skirting

Figure 6.18 Three types of panelling (*continued*)

On the job: Fitting door closers

The site agent asks Macy to fit two overhead door closers, giving her 30 minutes to finish them as the painter is coming to paint the door. Macy looks at the door closer and sees that, although she has fitted closers before, these are different. The site agent calls to Macy to hurry up and she fits the door closer the way she fitted the previous one. Macy is just tightening the last screw when the site agent comes over. He takes one look at the job and tells Macy that she has fitted them wrongly. Who is at fault? What could have been done to prevent it? What are the cost implications?

FAQ

Why are external doors thicker than internal doors?

They need to be more secure to prevent people breaking in.

Do I need to have a letter plate on my front door?

No. There are secure letter boxes that can be attached to the wall.

Do I have to use security hinges?

No, but security hinges are preferable as they prevent the door from being forced at the hinge side.

Why can't I have more than two deadlocks on a single exterior door?

Any more than two deadlocks would weaken the door at the lock side, as you will have removed a lot of material when mortising the locks in.

What type of door would I fit a security chain on?

Any door that doesn't have a glass panel or spy hole in it.

What is the best door closer to use?

The choice of door closer depends on the client's preference, but usually internal doors have concealed closers while doors in corridors have overhead closers.

Can I choose to use a dado profile for a picture rail?

If it is for decoration only, yes, but if it is to support paintings, no, because profiles for dado are not suitable for this.

Knowledge check

1. Name the four main types/styles of exterior door.

2. Give an example of where a framed, ledged and braced door would be used.

3. How many hinges are used when hanging a stable door?

4. On an external door, what is the purpose of a water bar?

5. Name five pieces of ironmongery that may be fitted on an external door.

6. State the size, type and number of hinges usually fitted to an external door.

7. What is the purpose of a spy hole?

8. What is the purpose of push plates?

9. Name the four main types of door closer.

10. What is the difference between standard double doors and double swing doors?

11. Name five different types of moulding that can be fitted in a room.

12. Give an example of where wall panelling can be found.

chapter 7

Structural carcassing

OVERVIEW

Structural carcassing covers all carpentry work associated with the structural elements of a building such as floors and roofs. This chapter is designed to help you identify the main activities associated with structural carcassing and to provide you with the knowledge and understanding required to carry them out.

It will cover the following:

- basic terms – a refresher

- traditional pitched roofs

- flat roofs

- ground and upper floors.

Basic terms – a refresher

Before looking at new and more complex areas of roofing, you should remind yourself of some of the terms and ideas you came across at Level 2. We will recap the key terms briefly now.

Roofing terminology

Roofs are made up of a number of different parts called 'elements'. These in turn are made up of 'members' or 'components'.

Elements

The main elements are defined below and shown in Figure 7.1.

- **gable** – the triangular part of the end wall of a building that has a pitched roof

- **hip** – where two external sloping surfaces meet

- **valley** – where two internal sloping surfaces meet

- **verge** – where the roof overhangs at the gable

- **eaves** – the lowest part of the roof surface where it meets the outside walls.

Members or components

The main members or components are defined below and shown in Figure 7.1.

- **ridge board** – a horizontal board at the apex acting as a spine, against which most of the rafters are fixed

- **wall plate** – a length of timber placed on top of the brickwork to spread the load of the roof through the outside walls and give a fixing point for the bottom of the rafters

- **rafter** – a piece of timber that forms the roof, of which there are several types

- **common rafters** – the main load-bearing timbers of the roof

- **hip rafters** – used where two sloping surfaces meet at an external angle, this provides a fixing for the jack rafters and transfers their load to the wall

- **crown rafter** – the centre rafter in a hip end that transfers the load to the wall

- **jack rafters** – these span from the wall plate to the hip rafter, enclosing the gaps between common and hip rafters, and crown and hip rafters

- **valley rafters** – like hip rafters but forming an internal angle, acting as a spine for fixing cripple rafters

- **cripple rafters** – similar to a jack rafter, these enclose the gap between the common and valley rafters

- **purlins** – horizontal beams that support the rafters mid-way between the ridge and wall plate.

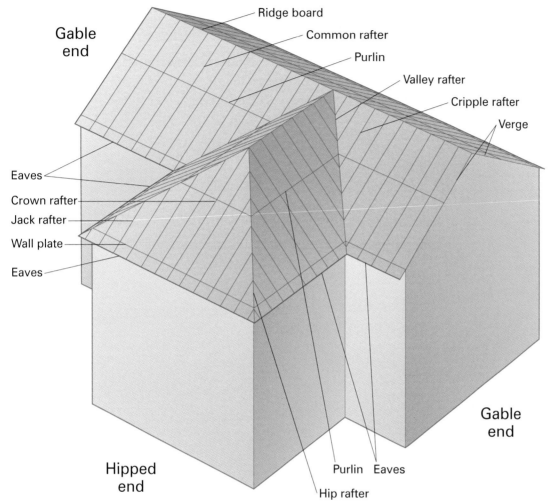

Figure 7.1 Roofing terminology

Basic terms for setting out

Setting out covers all the action necessary before commencing construction. Here are the most common terms you will need, which were covered in *Carpentry and Joinery NVQ and Technical Certificate Level 2* pages 241–247.

- **span** – the distance measured in the direction of the ceiling joists, from the outside of the wall plate to wall plate, known as the overall span

- **run** – equal to half the span

- **apex** – the peak or highest part of the roof

- **rise** – the distance from the outside of the wall plates at wall-plate level to the apex

- **pitch** – the angle or slope of the roof, calculated from the rise and the run

- **pitch line** – a line that is marked up from the underside of the rafter, one third of its depth to the top of the birdsmouth cut

- **plumb cut** – the angle of cut at the top of the rafter

- **seat cut** – the angle of cut at the bottom of the rafter

- **birdsmouth** – notch cut out at the bottom of the rafter to allow the rafter to sit on the wall plate.

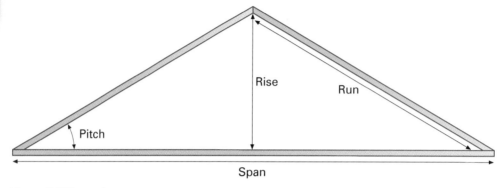

Figure 7.2 Rise and span

Traditional pitched roofs

There are several different types of **pitched roof** but most are constructed in one of two ways.

1. Trussed roof

A prefabricated pitched roof specially manufactured prior to delivery on site, saving timber as well as making the process easier and quicker. Trussed roofs can also span greater distances without the need for support from intermediate walls.

Trussed roofing is covered in *Carpentry and Joinery NVQ and Technical Certificate Level 2* page 271, so in Level 3 we will concentrate on traditional roofing.

2. Traditional roof

A roof entirely constructed on site from loose timber sections using simple jointing methods.

Roof types

A pitched roof can be constructed either as a single roof, where the rafters do not require any intermediate support, or a double roof where the rafters are supported. Single roofs are used over a short **span** such as a garage; double roofs are used to span a longer distance such as a house or factory.

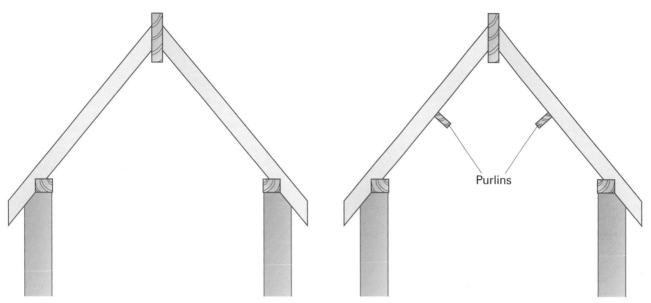

Figure 7.3 Single roof

Figure 7.4 Double roof

There are many different types of pitched roof including:

- **mono pitch** with a single pitch

- **lean-to** with a single pitch, which butts up to an existing building

- **duo pitch** with **gable ends**

- **hipped roof** with hip ends incorporating crown, hip and jack rafters

- **over hip** with gable ends, hips and valleys incorporating valley and cripple rafters

- **mansard** with gable ends and two different pitches used mainly when the roof space is to be used as a room

- **gable hip** or **gambrel** – double-pitched roof with a small gable (gablet) at the ridge and the lower part a half-hip

- **jerkin-head** or **barn hip** – double-pitched roof hipped from the ridge part-way to the eaves, with the remainder gabled.

The type of roof used will be selected by the client and architect.

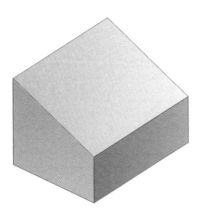

Figure 7.5 Mono pitch roof

Figure 7.6 Lean-to roof

Figure 7.7 Duo pitch roof with gable ends

Figure 7.8 Hipped roof

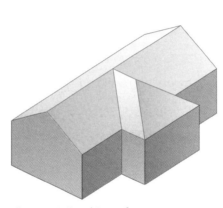

Figure 7.9 Over hip roof

Figure 7.10 Mansard roof

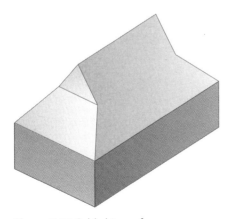

Figure 7.11 Gable hip roof

Figure 7.12 Jerkin-head roof

Gable ends

Setting out for a gable end

First set out and fix the wall plate. The wall plate is set on the brick or block work and either bedded in by the bricklayer or temporarily fixed by nailing through the joints. Once secured it is held in place with **restraint straps** (see Figure 7.13). If the wall plate is to be joined in length, a **halving joint** is used (see Figure 7.14). It is vital that the wall plate is fixed level to avoid serious problems later.

Safety tip

Using concrete or cement can cause dermatitis (an irritating skin complaint) so when working near wet cement be sure to wear proper gloves

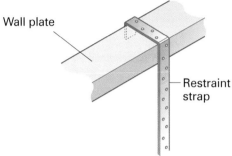

Figure 7.13 Restraint straps

Figure 7.14 Plate with lengthening joint

Once the wall plate is in place, you need to measure the span and the rise. You can use these measurements to work out the rafter length in different ways, using a **roofing ready reckoner**, **geometry** or **scale drawings**. Ready reckoners and geometry are covered later in this chapter, so we will start with scale drawings.

For this example we will use a span of 5m and a rise of 2.3m.

Using a scrap piece of plywood or hardboard we first draw the roof to a scale that will fit the scrap piece of plywood/hardboard (usually a scale of 1:20).

From this drawing we can measure at scale and find the **true length** of the rafter. Then by using a **sliding bevel** we can work out the plumb cut and seat cut.

Definition

Geometry – a form of mathematics using formulas, arithmetic and angles

Scale drawing – a drawing of a building or component in the right proportions, but scaled down to fit on a piece of paper. On a drawing at a scale of 1:50, a line 10mm long would represent 500mm on the actual object

Sliding bevel – a tool that can be set so that the user can mark out any angle

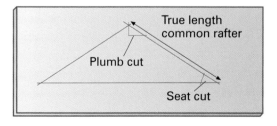

Figure 7.15 Sketch on a piece of scrap ply showing the true length, plumb and seat cut for a common rafter

Making and using a pattern rafter

From our scale drawing we can mark out one rafter, which we will then use as a **pattern rafter**.

There are five easy steps to follow when marking out a pattern rafter.

Step 1 Mark the pitch line one-third of the way up the width of the rafter.

Step 2 Set the sliding bevel to the plumb cut and mark the angle onto the top of the rafter.

Step 3 Mark the true length on the rafter, measuring along the pitch line.

Step 4 Use the sliding bevel to mark out the seat cut, then with a combination square mark out the birdsmouth at 90 degrees to the seat cut.

Step 5 Re-mark the plumb cut to allow for half the thickness of the ridge.

Once it has all been marked out, this can be cut and used as a pattern rafter.

The pattern rafter can be used to mark out all the remaining common rafters, although it is advisable to mark out and cut only four, then place two at each end of the roof to check whether the roof is going to be level.

Once all the rafters are cut, mark out the wall plate and fix the rafters. Rafters are normally placed at 400mm centres, with the first and last rafter 50mm away from the gable wall. The rafters are usually fixed at the foot by skew-nailing into the wall plate and at the head by nailing through the ridge board.

Did you know?

The first and last rafters are placed 50mm away from the wall to prevent moisture that penetrates the outside wall coming into contact with the rafters, thus preventing rot

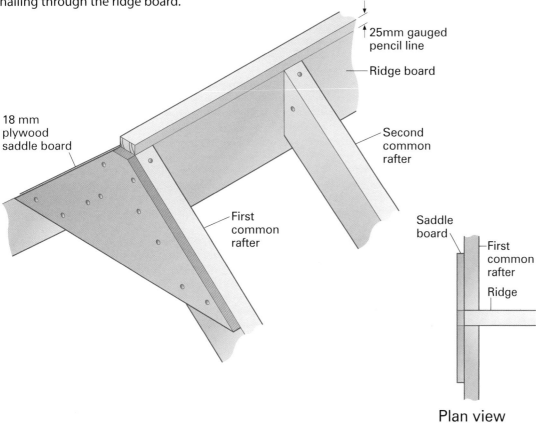

Figure 7.16 First two rafters fixed into place

If the roof requires a loft space, joists can be put in place and bolted to the rafters; if additional support is required, struts can be used.

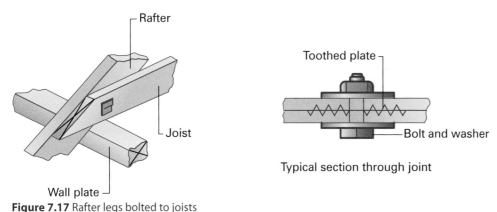

Figure 7.17 Rafter legs bolted to joists

For roofs with a large span, purlins provide adequate support. Purlins are usually built into the brickwork: either the gable wall will not have been finished being built yet or the bricklayer will have left a **sand course**. The same is true of roof ladders (see below).

Finishing a gable end

There are two types of finish for a gable end:

- a flush finish, where the bargeboard is fixed directly onto the gable wall
- a roof ladder – a frame built to give an overhang and to which the bargeboard and soffit are fixed.

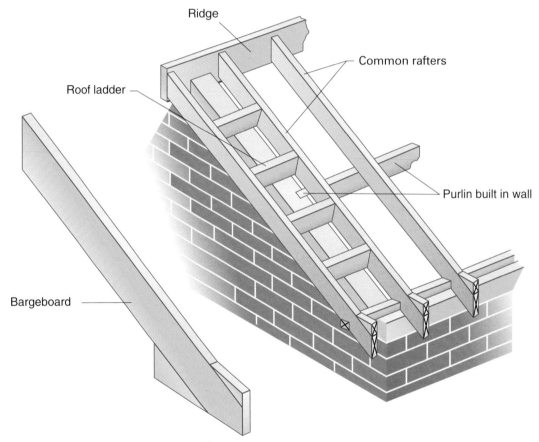

Figure 7.18 Roof ladder with bargeboard fitted

The most common way is to use a roof ladder, which when creating an overhang, stops rainwater running down the face of the gable wall.

The continuation of the fascia board around the verge of the roof is called the bargeboard. Usually the bargeboard is fixed to the roof ladder and has a built-up section at the bottom to encase the wall plate.

The simplest way of marking out the bargeboard is to temporarily fix it in place and use a level to mark the plumb and seat cut.

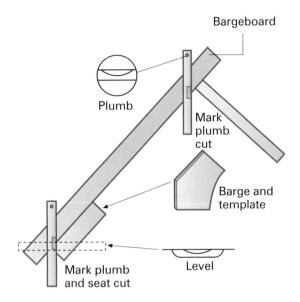

Figure 7.19 Marking out a bargeboard using a level

When fixing a bargeboard, the foot of the board may be mitred to the fascia, butted and finished flush with the fascia, or butted and extended slightly in front of the fascia to break the joint.

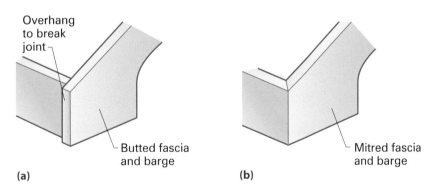

Figure 7.20 Fascia joined to bargeboard

The bargeboard should be fixed using oval nails or lost heads at least 2.5 times the thickness of the board so that a strong fixing is obtained. If there is to be a joint along the length of the bargeboard, the joint *must* be a mitre.

Hipped roofs

In a fully hipped roof there are no gables and the eaves run around the perimeter, so there is no roof ladder or bargeboard.

Marking out for a hipped roof

All bevels or angles cut on a hipped roof are based on the right-angled triangle and the roof members can be set out using the following two methods:

- **roofing ready reckoner** – a book that lists in table form all the angles and lengths of the various rafters for any span or rise of roof

- **geometry** – working with scale drawings and basic mathematic principles to give you the lengths and angles of all rafters.

The ready reckoner will be looked at later in this chapter, so for now we will concentrate on geometry.

Pythagoras' theorem

When setting out a hipped roof, you need to know Pythagoras' theorem. Pythagoras states that 'the square on the hypotenuse of a right-angled triangle is equal to the sum of the squares on the other two sides'. For the carpenter, the 'hypotenuse' is the rafter length, while the 'other two sides' are the run and the rise.

From Pythagoras' theorem, we get this calculation:

$A = \sqrt{B^2 + C^2}$ ($\sqrt{}$ means the square root and ^{2} means squared)

If we again look at our right-angled triangle we can break it down to:

A (the rafter length – the distance we want to know)

B^2 (the rise, multiplied by itself)

C^2 (the run, multiplied by itself).

Therefore we have all we need to find out the length of our rafter (A):

$A = \sqrt{4^2 + 3^2}$

$A = \sqrt{4 \times 4 + 3 \times 3}$

$A = \sqrt{16 + 9}$

$A = \sqrt{25}$

$A = 5$

so our rafter would be 5m long.

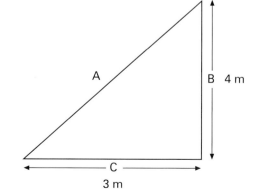

> ### Did you know?
>
> The three angles in a triangle always add up to 180 degrees

Finding true lengths

The next task is to find the hip rafter true length, plumb and seat cuts. This is done in two stages (the first will be familiar to you, as it is the same as for a common rafter).

The next step is to lengthen the common rafter true length by the amount of the rise, then join this line up to the base of the roof.

From this point, to make the geometrical drawings as clear as possible, abbreviated labels will be used. See Table 7.1.

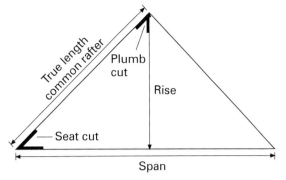

Figure 7.21 Finding common rafter true length

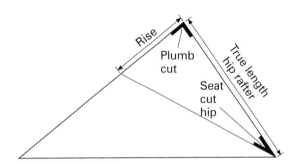

Figure 7.22 Finding hip rafter true length

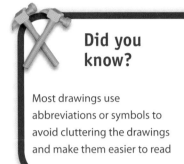

Did you know?

Most drawings use abbreviations or symbols to avoid cluttering the drawings and make them easier to read

Abbreviation	Definition	Abbreviation	Definition
TL	true length	PCHR	plumb cut hip rafter
TLCR	true length common rafter	SCHR	seat cut hip rafter
TLHR	true length hip rafter	PCVR	plumb cut valley rafter
TLJR	true length jack rafter	SCVR	seat cut valley rafter
TLCrR	true length cripple rafter	EC	edge cut
PC	plumb cut	ECHR	edge cut hip rafter
SC	seat cut	ECJR	edge cut jack rafter
PCCR	plumb cut common rafter	ECVR	edge cut valley rafter
SCCR	seat cut common rafter	ECCrR	edge cut cripple rafter

Table 7.1 Abbreviations

There are two other angles that are concerned with hip rafters: the dihedral angle (or backing bevel for the hip) and the edge cut to the hip.

Finding the dihedral or backing bevel angle

The backing bevel angle is the angle between the two sloping roof surfaces. It provides a level surface so that the tile battens or roof boards can lie flat over the hip rafters. The backing bevel angle is rarely used in roofing today as the edge of the hip is usually worked square, but you should still know how to work it out.

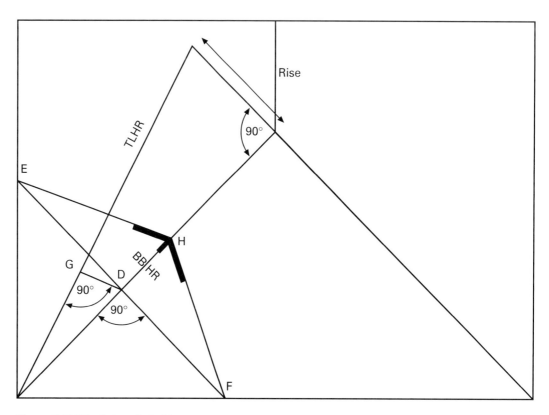

Figure 7.23 Dihedral angle for hip

1. Draw a plan of the roof and mark on the TLHR as before.

2. Draw a line at right angles to the hip on the plan at D, to touch the wall plates at E and F.

3. Draw a line at right angles to the TLHR at G, to touch point D.

4. With centre D and radius DG, draw an arc to touch the hip at H.

5. Join E to H and H to F. This gives the required backing bevel (BBHR).

Finding the edge cut

The edge cut is applied to both sides of the hip rafter at the plumb cut. It enables the hip to fit up to the ridge board between the crown and the common rafters.

1. Draw a plan of the roof and mark on the TLHR as before.

2. With centre I and radius IB, swing the TLHR down to J, making IJ, TLHR.

3. Draw lines at right angles from the ends of the hips and extend the ridge line. All three lines will intersect at K.

4. Join K to J. Angle IJK is the required edge cut (ECHR).

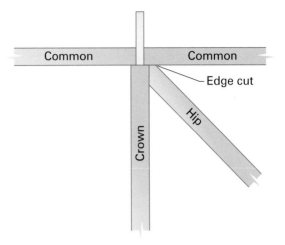

Figure 7.24 Edge cut on hip joined to ridge

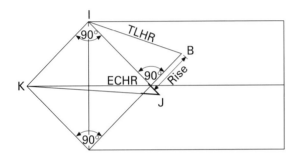

Figure 7.25 Edge cut on hip

The jack rafter's plumb and seat cuts are the same as those used in the common rafters so all you need to work out is the true length and edge cut.

*   Draw the plan and section of the roof. Mark on the plan the jack rafters. Develop roof surfaces by swinging TLCR down to L and projecting down to M^1

*   With centre N and radius NM^1, draw arc M^1O. Join points M^1 and O to ends of hips as shown.

*   Continue jack rafters on to development.

*   Mark the true length of jack rafter (TLJR) and edge cut for jack rafter (ECJR).

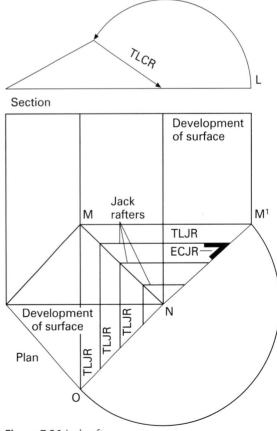

Figure 7.26 Jack rafter length and cuts

Setting out a hipped end

First we need to fit the wall plate, then the ridge and common rafters. To know where to place the common rafters you need to work out the true length of the rafter, then begin to mark out the wall plate.

The wall plate is joined at the corners and marked out as shown if figure 7.28.

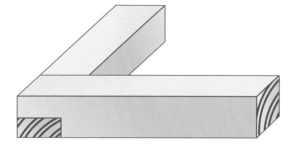

Figure 7.27 Corner halving

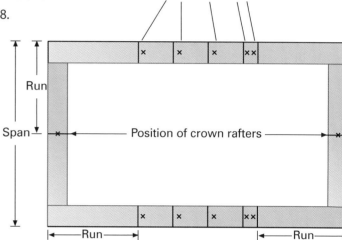

Figure 7.28 Wall plate marked out

Marking out for a hipped roof

To mark out for a hipped roof, follow these steps:

1. Measure the span and divide it by two to get the run, then mark this on the hipped ends – the centre of your crown rafter will line up with this mark.

2. Mark the run along the two longer wall plates – these marks will give you the position of your first and last common rafter.

3. The common rafter will sit to the side of this line, so a cross or other mark should be made to let you know which side of the line the rafter will sit.

4. Mark positions for the rest of the rafters on the wall plate at the required centres, again using a cross or other mark to show you which side of the line the rafter will sit. Note: the last two rafters may be closer together than the required centres but must not be wider apart than the required centres.

5. Cut and fit the common rafters using the same method as used for a gable roof.

6. Fit the crown rafter, which has the same plumb and seat cuts as the common rafter and is almost the same length – but here you should *not* remove half the ridge thickness.

Definition

Crown rafter – the rafter that sits in the middle of a hip end, between the two hips

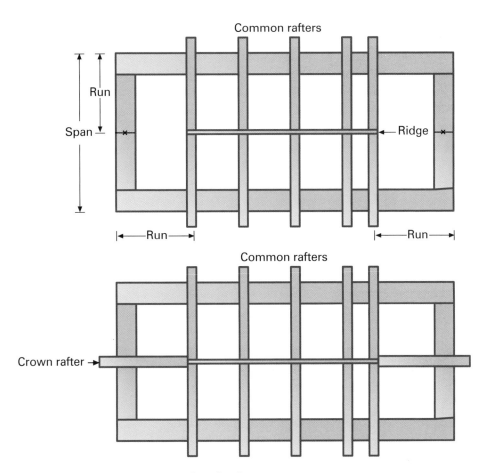

Figure 7.29 Common and crown rafters fitted

7. For the hip rafters, work out the true length, all the angles and bevels and mark out one hip as shown below, then cut the hip and try it in the four corners. If the hip fits in all four corners, you can use it as a template to mark out the rest of the hips; if not, the roof is **out of square** or level, but you can still use this hip to help mark out the remaining three corners.

Figure 7.30 Pitch line marked on hip

With a hip rafter it is important to remember that the pitch line is marked out differently. It is marked from the top of the rafter and is set at two-thirds of the depth of the *common* rafter.

The best way to check your rafter before cutting is to measure from the point of the ridge down to the corner of the wall plate. This distance should be the same as marked on your rafter.

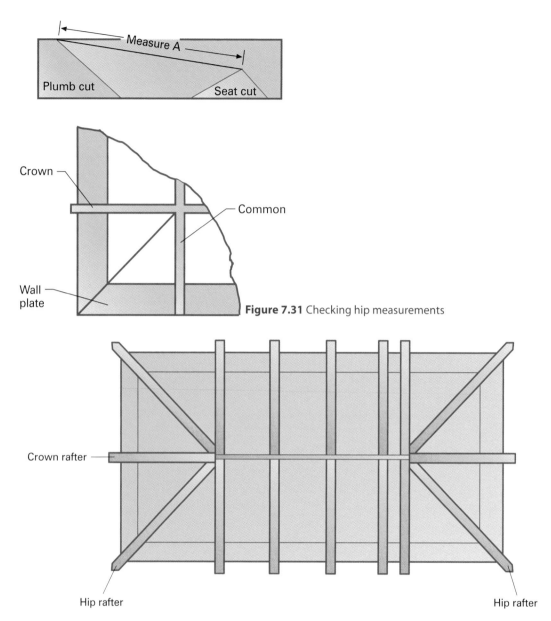

Figure 7.31 Checking hip measurements

Figure 7.32 Hip rafters in place

8. Cut in the jack rafters. Find out the true length and edge cut of all the jack rafters, then mark them out and cut them. As the jack rafters are of different sizes it is better to cut them individually to fit. They can still be used as template rafters on the opposite side of the hip.

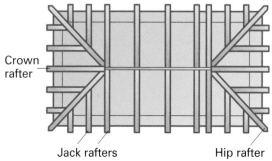

Figure 7.33 Jacks fitted

Valleys

The next section deals with valleys, which are formed when two sloping parts of a pitched roof meet at an internal corner.

Marking out for valleys

Valleys can be worked out in the same way as hips, using either a ready reckoner or geometry. Here we will look at geometry.

Working out the angles for valleys is similar to doing so for hips except that the key drawing is not a triangle but a plan drawing of the roof.

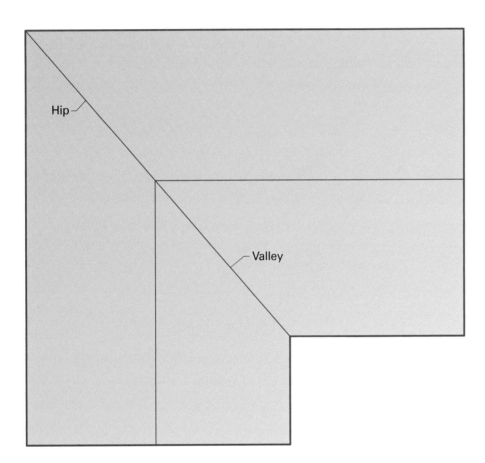

Figure 7.34 Plan of roof

First you need to find out the valley rafter true length, plumb and seat cut. Start by finding the rise of the roof and drawing a line this length at a right angle to the valley where it meets the ridge. Join this line to the point where the valley meets the wall plate. This will give you the true length of the valley rafter as well as the plumb and seat cuts.

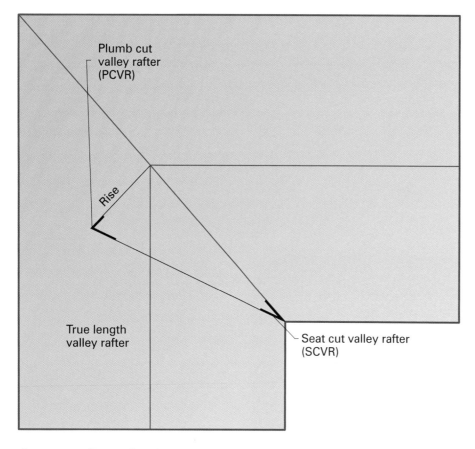

Figure 7.35 Valley true length

As with the hip rafter there are two other angles to find for a valley rafter: the dihedral angle and the edge cut.

The dihedral angle for the valley is used in the same way as the hip dihedral and again rarely in roofing today. Figure 7.36 shows you how to work out the dihedral angle.

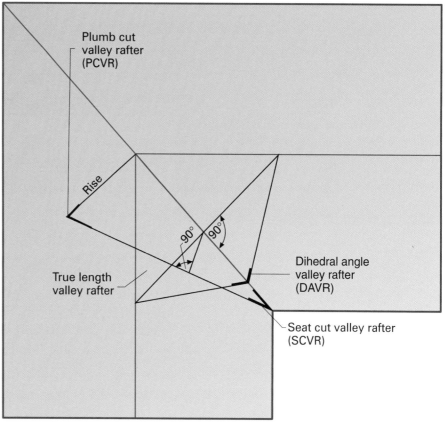

Figure 7.36 Dihedral angle hip

The final angle to find is the edge cut for the valley rafter, as follows.

1. Mark on the rise and true length of the valley rafter.

2. Draw a line at right angles to the valley where it meets the wall plate and extend this line to touch the ridge at A.

3. Set your compass to the true length of the valley and **swing an arc** towards the ridge at B.

4. Join up the line A–B to give you the edge cut.

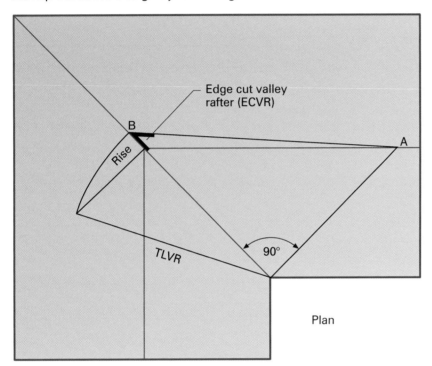

Plan

Figure 7.37 Edge cut

The final part of valley geometry is to find the true length and edge cut for the cripple rafters, as follows.

1. Draw out the roof plan as usual, then to the side of your plan draw out a section of the roof.

2. Set your compass to the rafter length and swing an arc downwards.

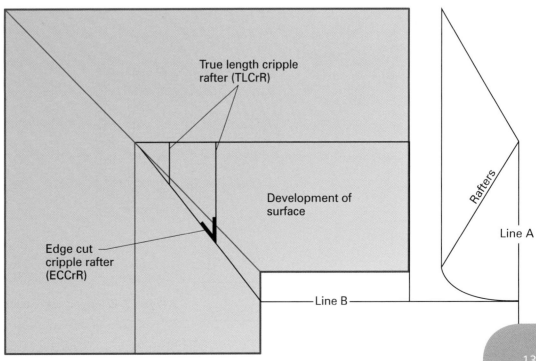

Figure 7.38 Cripple length and angle

3. Draw line A downwards until it meets the arc, then draw a line at right angles to line A until it hits the wall plate, creating line B.

4. Draw a line from where line B hits the wall plate up to where the valley meets the ridge. This will give you the appropriate true length (TLCrR) and edge cut (ECCrR).

Setting out a valley

There are four steps to follow when setting out a valley.

Step 1 Fit the wall plate and mark it out with the position of the common rafters.

Step 2 Fit the common rafters and ridge.

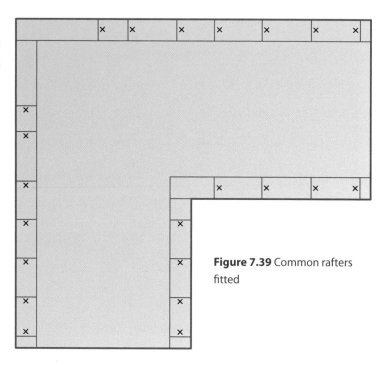

Figure 7.39 Common rafters fitted

Step 3 Fit the hip and jack rafters.

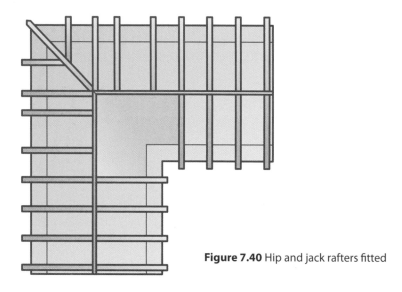

Figure 7.40 Hip and jack rafters fitted

Step 4 Fit the valley and cripple rafters, taking the true lengths and bevels from the drawings.

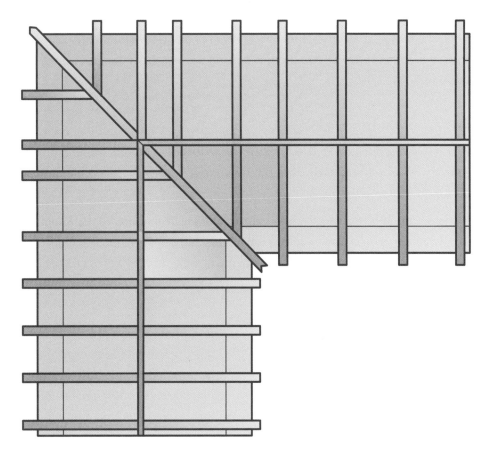

Figure 7.41 Valley and cripple rafters fitted

An alternative to using valley rafters is to use a lay board. **Lay boards** are most commonly used with extensions to existing roofs or where there are dormer windows.

The lay board is fitted onto the existing rafters at the correct pitch, then the cripple rafters are cut and fixed to it.

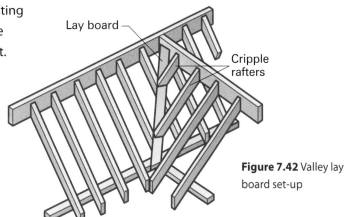

Lay board

Cripple rafters

Figure 7.42 Valley lay board set-up

Definition

Lay board – a piece of timber fitted to the common rafters of an existing roof to allow the cripple rafters to be fixed

Trimming, eaves and covering

Trimming roof openings

Roofs often have components such as chimneys or roof windows. These components create extra work, as the roof must have an opening for them to be fitted. This involves cutting out parts of the rafter and putting in extra support to carry the weight of the roof over the missing rafters.

Chimneys

Chimneys are rarely used in new house construction as there are more efficient and environmental ways of heating these days, but most older houses will have chimneys and these roofs must be altered to suit.

When constructing such a roof, the chimney should already be in place, so you should cut and fit the rest of the roof, leaving out the rafters where the chimney is. When you mark out the wall plate, make sure that the rafters are positioned with a 50mm gap between the chimney and the rafter. You may also need to put in extra rafters.

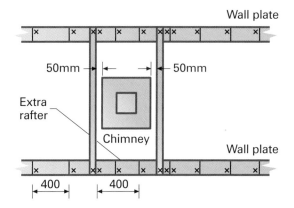

Figure 7.43 Wall plate marked out to allow for chimney

Next, fit the trimmer pieces between the rafters to bridge the gap, then fix the trimmed rafters – rafters running from wall plate to trimmer and from trimmer to ridge. The trimmed rafters are birdsmouthed at the bottom to sit over the wall plate, and the plumb and seat cut is the same as for common rafters.

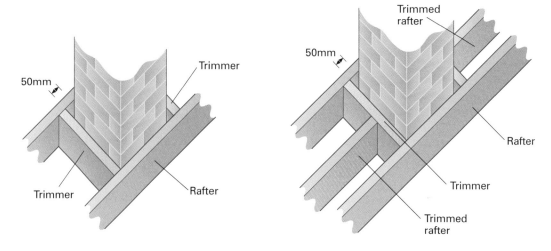

Figure 7.44 Opening trimmed around chimney with trimmers fitted

If the chimney is at **mid-pitch** rather than at the ridge, you will need to fit a chimney gutter: this ensures the roof remains watertight by preventing the water gathering at one point. The chimney gutter should be fixed at the back of the stack.

Definition

Mid-pitch – in the middle of the pitch rather than at the apex or eaves

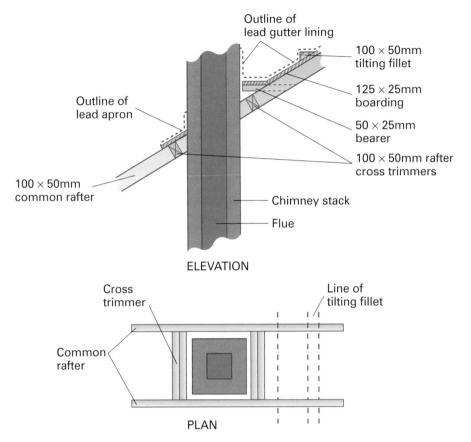

Outline of lead gutter lining

100 × 50mm tilting fillet

125 × 25mm boarding

Outline of lead apron

50 × 25mm bearer

100 × 50mm rafter cross trimmers

100 × 50mm common rafter

Chimney stack

Flue

ELEVATION

Cross trimmer

Line of tilting fillet

Common rafter

PLAN

Figure 7.45 Gutter detail around chimney

Roof windows and skylights

You need to trim openings for roof windows and skylights too.

If the roof is new, you can plan it in the same way as for a chimney, remembering to make sure the area you leave for trimming is the same size as the window.

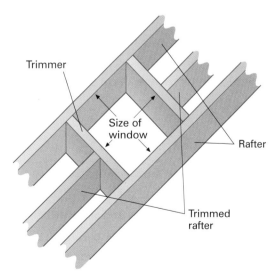

Trimmer

Size of window

Rafter

Trimmed rafter

Figure 7.46 Roof trimmed for roof light

If you need to fit a roof window in an existing roof the procedure is different.

1. Strip all the tiles or slates from the area where you want to fit the window.

2. Remove the tile battens, felt and roof cladding.

3. Mark out the position of the window and cut the rafters to suit.

4. Now fit the trimmer and trimmed rafters– you may need to double up the rafters for additional strength – and fit the window.

5. Re-fit the cladding and felt, and fix any flashings.

6. Cut and fit new tile battens and finally re-tile or re-slate the roof.

Dormers

Dormer windows are different to roof windows: a roof window lays flat against the roof while a dormer window projects up from the sloping surface. A dormer window is often preferred to a roof window when there is limited headroom.

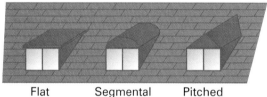

Flat   Segmental   Pitched

Figure 7.47 Dormer types

The construction of a dormer is similar to that of a roof window, except once the opening is trimmed, you need to add a framework to give the dormer shape. For a dormer you usually double up the rafters on either side to give support for the extra load the dormer puts on the roof.

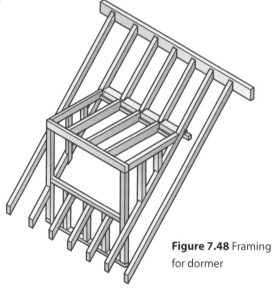

Figure 7.48 Framing for dormer

Eaves details

The eaves are how the lower part of the roof is finished where it meets the wall, and incorporates fascia and soffit. The fascia is the vertical board fixed to the ends of the rafters. It is used to close the eaves and allow fixing for rainwater pipes. The soffit is the horizontal board fixed to the bottom of the rafters and the wall. It is used to close the roof space to prevent birds or insects from nesting there, and usually incorporates ventilation to help prevent rot.

There are various ways of finishing a roof at the eaves; we will look at the four most common.

Flush eaves

Here the eaves are finished as close to the wall as possible. There is no soffit, but a small gap is left for ventilation.

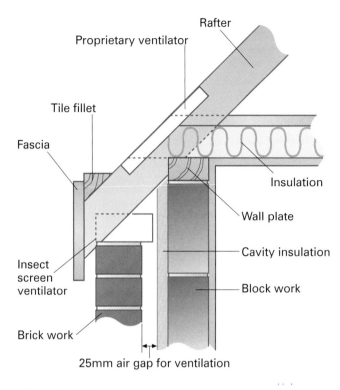

Figure 7.49 Flush eaves

Open eaves

An open eaves is where the bottom of the rafter feet are planed as they are exposed. The rafter feet project beyond the outer wall and eaves boards are fitted to the top of the rafters to hide the underside of the roof cladding. The rainwater pipes are fitted via brackets fixed to the rafter ends.

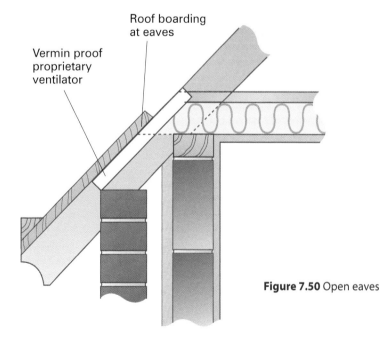

Figure 7.50 Open eaves

143

Closed eaves

Closed eaves are completely closed or boxed in. The ends of the rafters are cut to allow the fascia and soffit to be fitted. The roof is ventilated either by ventilation strips incorporated into the soffit or by holes drilled into the soffit with insect-proof netting over them. If closed eaves are to be re-clad due to rot you must ensure that the ventilation areas are not covered up.

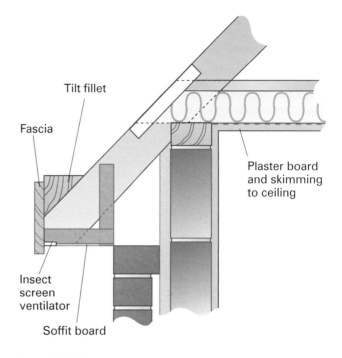

Figure 7.51 Closed eaves

Definition

Sprocket – a piece of timber bolted to the side of the rafter to reduce the pitch at the eaves

Sprocketed eaves

Sprocketed eaves are used where the roof has a sharp pitch. The **sprocket** reduces the pitch at the eaves, slowing down the flow of rainwater and stopping it overshooting the guttering. Sprockets can either be fixed to the top edge of the rafter or bolted onto the side.

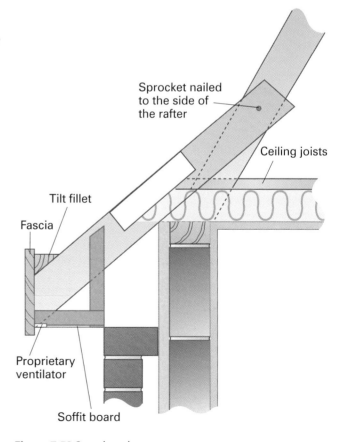

Figure 7.52 Sprocketed eaves

Roof coverings

Once all the rafters are on the roof, the final thing is to cover it. There are two main methods of covering a roof, each using different components. Factors affecting the choice of roof covering include what the local weather is like and what load the roof will have to take.

Method 1

This method is usually used in the north of the country where the roof may be expected to take additional weight from snow.

1. Clad the roof surface with a man-made board such as OSB or exterior grade plywood.

2. Cover the roof with roofing felt starting at the bottom and ensuring the felt is overlapped to stop water getting in.

3. Fit the felt battens (battens fixed vertically and placed to keep the felt down while allowing ventilation) and the tile battens (battens fixed horizontally and accurately spaced to allow the tiles to be fitted with the correct overlap).

4. Finally, fit the tiles and cement on the ridge.

Method 2

This is the most common way of covering a roof.

1. Fit felt directly onto the rafters.

2. Fit the tile battens at the correct spacing.

3. Fit the tiles and cement on the ridge.

Another way to cover a roof involves using slate instead of tiles. Slate-covered roofing is a specialised job as the slates often have to be cut to fit, so roofers usually carry this out.

Ready reckoner

A ready reckoner is a book used as an alternative to the geometry method and is often the simplest way of working out lengths and bevels. The book consists of a series of tables that are easy to follow once you understand the basics.

To use the ready reckoner you must know the span and the pitch of the roof.

Did you know?

Where you live may have an affect on your choice of roof: in areas more prone to bad weather, the roof will need to be stronger

Example

Take a hipped roof with a 36 degree pitch and a span of 8.46m. First you halve the span, getting a run of 4.23m. Referring to the tables in the ready reckoner, you can work out the lengths of the common rafter as follows:

RISE OF COMMON RAFTER 0.727m PER METRE RUN   PITCH 36 degrees

BEVELS:	COMMON RAFTER	SEAT 36
	COMMON RAFTER	PLUMB 54
	HIP OR VALLEY	SEAT 27
	HIP OR VALLEY	RIDGE 63
	JACK RAFTER	EDGE 39

JACK RAFTERS	333mm CENTRES DECREASE	412mm
	400mm CENTRES DECREASE	494mm
	500mm CENTRES DECREASE	618mm
	600mm CENTRES DECREASE	742mm

Run of Rafter	0.1	0.2	0.3	0.4	0.5	0.6	0.7	0.8	0.9	1.0
Length of Rafter	0.124	0.247	0.371	0.494	0.618	0.742	0.865	0.989	1.112	1.236
Length of Hip	0.159	0.318	0.477	0.636	0.795	0.954	1.113	1.272	1.431	1.590

You can see that the length of a rafter for a run of 0.1m is 0.124m, therefore for a run of 1m, the rafter length will be 1.24m – but you need to find the length of a rafter for a run of 4.23m. This is how:

        1.00m = 1.24m

So: 4.00m = 4.994m

        0.20m = 0.247m

        0.03m = 0.037m

        total   = 5.228m

So the length of the common rafter is 5.228m. However there are a few adjustments you must make before finding the finished size.

You need to allow for an overhang and for the ridge, which can both easily be measured. For the purposes of this example, we will use an overhang of 556mm and a ridge of 50mm. The final calculation is:

Basic rafter   5.228m

+ overhang   0.556m

– half ridge   0.025m

total   **5.759m**

So the rafter length is 5.759m.

Now you need to refer back to the table, which tells you that, for the common rafter, the seat cut is 36 degrees and the plumb cut is 54 degrees.

You can now mark out and cut your pattern rafter as before, remembering to mark on the pitch line and plumb cut. Measure the original size (5.228m for our example) along the plumb cut, mark out the seat cut and finally cut.

The hip and valley rafters are worked out in the same way but jack and cripple rafters are different.

Jack and cripple rafters are marked on the table and are easy to work out. Continuing with the example, where the rafter length is 5.228m, you look at the table and, if you are working at spacing the rafters at 400mm centres, you reduce the length of the common rafter by 494mm. The first jack/cripple rafter will be 5.228m – 0.494m = 4.734m; the next jack/cripple rafter will be 4.734m – 0.494m = 4.240m and so on. The angles for all the rest of the cuts are all shown on the table.

BEVELS:	COMMON RAFTER	SEAT 36
	COMMON RAFTER	PLUMB 54
	HIP OR VALLEY	SEAT 27
	HIP OR VALLEY	RIDGE 63
	JACK RAFTER	EDGE 39

JACK RAFTERS	333mm CENTRES DECREASE	412mm
	400mm CENTRES DECREASE	494mm
	500mm CENTRES DECREASE	618mm
	600mm CENTRES DECREASE	742mm

Remember

The plumb and seat cuts for the jack rafter are the same as for the common rafter

Flat roofs

A flat roof is any roof which has its upper surface inclined at an angle (also known as the fall, slope or pitch) not exceeding 10 degrees.

A flat roof has a fall to allow rainwater to run off, preventing puddles forming as they can put extra weight on the roof and cause leaks. Flat roofs will eventually leak, so most are guaranteed for only 10 years (every 10 years or so the roof will have to be stripped back and re-covered). Today **fibreglass** flat roofs are available that last much longer, so some companies will give a 25-year guarantee on their roof. Installing a fibreglass roof is a job for specialist roofers.

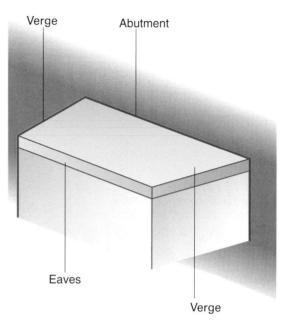

Figure 7.53 Flat roof terminology

The amount of fall should be sufficient to clear water away to the outlet pipe(s) or guttering as quickly as possible across the whole roof surface. This may involve a single direction of fall or several directional changes of fall such as:

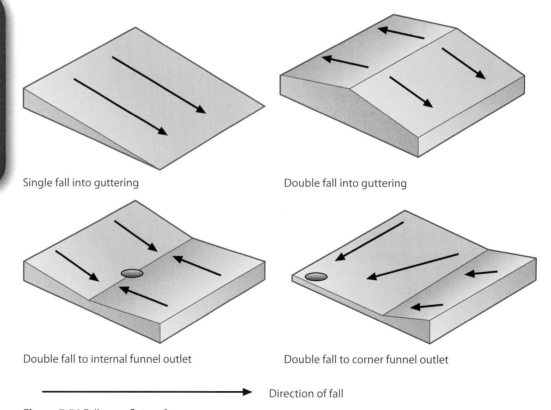

Single fall into guttering

Double fall into guttering

Double fall to internal funnel outlet

Double fall to corner funnel outlet

→ Direction of fall

Figure 7.54 Falls on a flat roof

Construction of a flat roof

Flat roofs joists are similar in construction to floor joists (discussed later in this chapter) but unless they are to be accessible, they are not so heavily loaded. Joists are, therefore, of a smaller dimension than those used in flooring.

There are many ways to provide a fall on a flat roof. The method you choose depends on what the direction of fall is and where on a building the roof situated.

Laying joists to a fall

This method is by far the easiest: all you have to do is ensure that the wall plate fixed to the wall is higher than the wall plate on the opposite wall or vice versa. The problem is that this method will also give the interior of the roof a sloping ceiling. This may be fine for a room such as a garage, but for a room such as a kitchen extension the client might not want the ceiling to be sloped and another method would have to be used.

Joists with firring pieces

Firring pieces provide a fall without disrupting the interior of the room, but involve more work. Firrings can be laid in two different ways, depending on the layout of the joists and the fall.

The lay out of joists is explained in more detail in the flooring section of this chapter.

> **Definition**
>
> **Firring piece** – long wedge tapered at one end and fixed on top of joists to create the fall on a flat roof

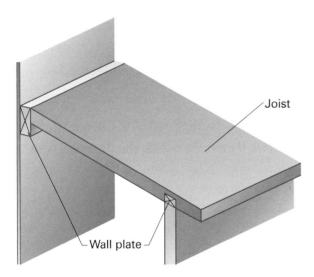

Figure 7.55 Joists laid to a fall

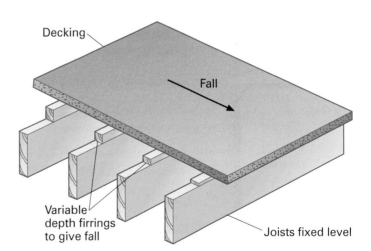

Figure 7.56 Joists with firring pieces

Using firring pieces

The basic construction of a flat roof with firrings begins with the building of the exterior walls. Once the walls are in place at the correct height and level, the carpenter fits the wall plate on the eaves wall (there is no need for the wall plate to be fitted to the verge walls). This can be bedded down with cement, or nailed through the joints in the brick or block work with restraining straps fitted for extra strength. The carpenter then fixes the **header** to the existing wall.

The header can either be the same depth as the joists or have a smaller depth. If it is the same depth, the whole of the joist butts up to the header and the joists are fixed using joist hangers; if the header has a smaller depth, the joists can be notched to sit on top of the header as well as using framing anchors.

Once the wall plate and header are fixed, they are marked out for the joists at the specified centres (300mm, 400mm or 600mm). The joists are cut to length, checked for **camber (crown)** and fixed in place using joist hangers. Strutting or **noggings** are then fitted to help strengthen the joists.

Once the joists are fixed in place they must have restraining straps fitted. A strap must be fitted to a minimum of one joist per 2m run, then firmly anchored to the wall to prevent movement in the roof under pressure from high winds.

The next step is to fix the firring pieces, which are either nailed or screwed down onto the top of the joists. Insulation is fitted between the joists, along with a vapour barrier to prevent the movement of moisture caused by condensation.

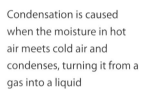

Decking

Once the insulation and vapour barrier are fitted, it is time to fit the decking. As you may remember from Level 2, decking a flat roof can be done with a range of materials including:

- **tongued and grooved board**

These boards are usually made from pine and are not very moisture-resistant, even when treated, so they are rarely used for decking these days. If used, the boards should be laid either with, or diagonal to, the fall of the roof. Cupping of the boards laid across the fall could cause the roof covering to form hollows in which puddles could form.

- **plywood**

Only roofing grade boards stamped with **WBP** (weather/water boil proof) should be used. Boards must be supported and securely fixed on all edges in case there is any distortion, which could rip or tear the felt covering. A 1mm gap must be left between each board along all edges in case there is any movement caused by moisture, which again could cause damage to the felt.

- **chipboard**

Only the types with the required water resistance classified for this purpose must be used. Boards are available that have a covering of bituminous felt bonded to one surface, giving temporary protection against wet weather. Once laid, the edges and joints can be sealed. Edge support, laying and fixing are similar to floors (covered later in this chapter). Moisture movement will be greater than with plywood as chipboard is more **porous**, so a 2mm gap should be allowed along all joints, with at least a 10mm gap around the roof edges. Tongued and grooved chipboard sheets should be fitted as per the manufacturer's instructions.

- **oriented strand board (OSB)**

Generally more stable than chipboard, but again only roofing grades must be used. Provision for moisture movement should be made as with chipboard.

- **cement-bonded chipboard**

Strong and durable with high density (much heavier and greater moisture resistance than standard chipboard). Provisions for moisture movement should be the same as chipboard.

- **metal decking**

Profiled sheets of aluminium or galvanised steel with a variety of factory-applied colour coatings and co-ordinated fixings are available. Metal decking is more usually associated with large steel sub-structures and fixed by specialist installers, but it can be used on small roof spans to some effect. Sheets can be rolled to different profiles and cut to any reasonable length to suit individual requirements.

- **translucent sheeting**

This might be corrugated or flat (e.g. polycarbonate twin wall), and must be installed as per the manufacturer's instructions.

Metal decking and translucent sheeting are supplied as finished products but timber-based decking needs additional work to it to make it watertight, as explained in the next section.

Weatherproofing a flat roof

Once a flat roof has been constructed and decked the next step is to make it watertight. The roof decking material can be covered using different methods and with different materials. One basic way of covering the decking is as follows.

The roof decking is covered in a layer of hot **bitumen** to seal any gaps in the joints. Then another layer of bitumen is poured over the top and felt is rolled onto it, sticking fast when the bitumen sets. A second roll of felt is stuck down with bitumen, but is laid at 90 degrees to the first. Some people add more felt at this stage – sometimes up to five more layers.

The final step can also be done in a number of ways. Some people put stones or chippings down on top of the final layer; some use felt that has stones or chippings imbedded and a layer

Definition

Porous – full of pores or holes: if a board is porous, it may soak up moisture or water and swell

Translucent – allowing some light through but not clear like glass

Bitumen – also known as pitch or tar, bitumen is a black sticky substance that turns into a liquid when heated, and is used on flat roofs to provide a waterproof seal

Did you know?

Bitumen is bought as a solid material. It is then broken up, placed in a cauldron and heated until it becomes a liquid, which can be spread over the flat roof. The bitumen will only stay in liquid form for a few minutes before it starts to set. Once set, it forms a waterproof barrier.

of dried bitumen on the back. The felt with stones imbedded into it is laid by rolling the felt out and using a gas blowtorch to heat the back, which softens the bitumen allowing it to stick.

Finishing a flat roof at the abutment and verge

Abutment

The abutment finish needs to take into consideration the existing wall as well as the flat roof. The abutment is finished by cutting a slot into the brick or block work and fixing lead to give a waterproof seal, which prevents water running down the face of the wall and into the room. **Tilt** or **angle fillets** are used to help with the run of the water and to give a less severe angle for the lead to be dressed to.

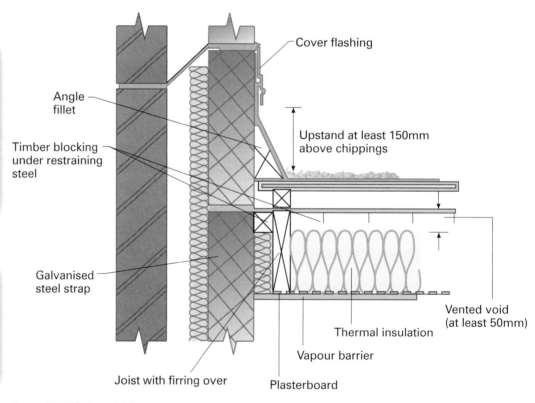

Cover flashing

Angle fillet

Timber blocking under restraining steel

Upstand at least 150mm above chippings

Galvanised steel strap

Vented void (at least 50mm)

Thermal insulation

Vapour barrier

Joist with firring over

Plasterboard

Figure 7.57 Abutment joint

Verge

Since a flat roof has such a shallow pitch, the verge needs some form of upstand to stop the water flowing over the sides at the verge instead of into the guttering at the eaves. This is done using tilt or angle fillets nailed to the decking down the full length of the verge prior to the roof being felted or finished.

Eaves details

The eaves details can be finished in the same way as the eaves on a pitched roof, with the soffit fitted to the underside of the joists and the fascia fitted to the ends.

Ground and upper floors

Several types of flooring are used in the construction of buildings, ranging from timber floors to large pre-cast concrete floors, which are used in large buildings such as residential flats. The main type of flooring a carpenter will deal with is suspended timber and that is what we will concentrate on in this section. A suspended timber floor can be fitted at any level from top floor to ground floor. In the next few pages we will look at:

- basic structure
- joists
- construction methods
- floor coverings.

Basic structure

Suspended timber floors are constructed with timbers known as joists, which are spaced parallel to each other spanning the distance of the building. Suspended timber floors are similar to traditional roofs in that they can be single or double, a single floor being supported at the two ends only and a double floor supported at the two ends and in the middle by way of a sleeper/dwarf wall, steel beam or load-bearing partition.

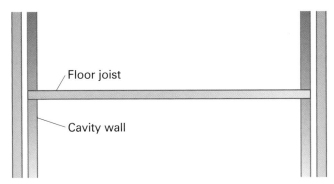

Floor joist

Cavity wall

Figure 7.58 Single floor

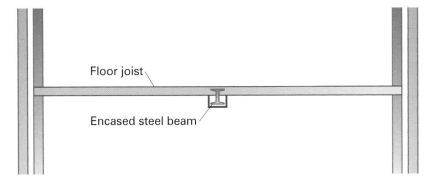

Floor joist

Encased steel beam

Figure 7.59 Double floor

All floors must be constructed to comply with Building Regulations, in particular Part C, which is concerned with damp. The bricklayer must insert a **damp proof course (DPC)** between the brick or block work when building the walls, situated no less than 150mm above ground level. This prevents moisture moving from the ground to the upper side of the floor. No timbers are allowed below the DPC. Air bricks, which are built into the external walls of the building, allow air to circulate round the underfloor area, keeping the moisture content below the dry rot limit of 20 per cent, thus preventing dry rot.

Joists

In domestic dwellings suspended upper floors are usually single floors, with the joist supported at each end by the structural walls but, if support is required, a load-bearing partition is used. The joists that span from one side of the building to the other are called bridging joists, but any joists that are affected by an opening in the floor such as a stairwell or chimney are called trimmer, trimming and trimmed joists.

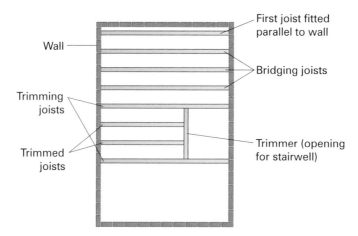

Figure 7.60 Joists and trimmers

Bridging joists are usually sawn timber 50mm thick. The trimmer that carries the trimmed joists and transfers this load to the trimming must be thicker – usually 75mm sawn timber, or in some instances two 50mm bridging joists bolted together. The depth of the joist is easily worked out by using the calculation:

Depth of joist = span / 20 + 20

So, for example, if you have a span of 4m

Depth = 4000/20 + 20

Depth = 200 + 20

Depth = 220

the depth of the joist required would be 220mm.

If the span was 8m, the depth would double to 440mm. A depth of 440mm is too great, so you would need to look at putting in a support to create a double floor.

Construction methods

A suspended timber floor must be supported either end. Figures 7.61 and 7.62 show ways of doing this.

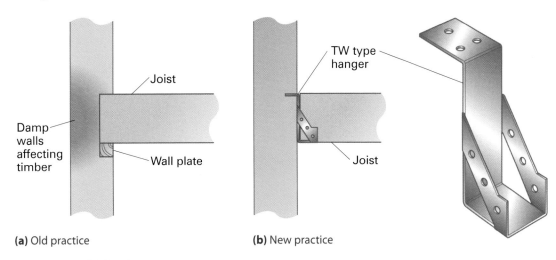

(a) Old practice

(b) New practice

Figure 7.61 Solid floor bearings

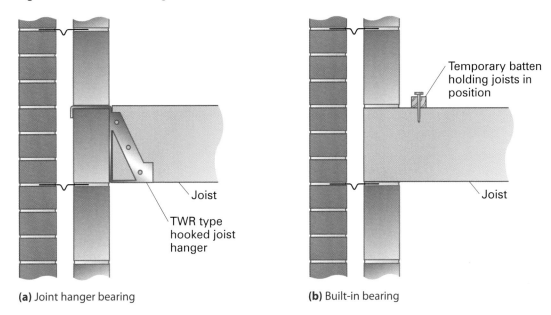

(a) Joint hanger bearing

(b) Built-in bearing

Figure 7.62 Cavity wall bearings

If a timber floor has to trim an opening, there must be a joint between the trimming and the trimmer joists. Traditionally, a **tusk tenon joint** was used (even now, this is sometimes preferred), between the trimming and the trimmer joist. If the joint is formed correctly, a tusk tenon is extremely strong, but making one is time-consuming. A more modern method is to use a metal framing anchor or timber-to-timber joist hanger.

Traditional tusk tenon joint

Joist hanger

Fitting floor joists

Before the carpenter can begin constructing the floor, the bricklayer needs to build the honeycomb sleeper walls. This type of walling has gaps in each course to allow the free flow of air through the underfloor area. It is on these sleeper walls that the carpenter lays his timber wall plate, which will provide a fixing for the floor joists.

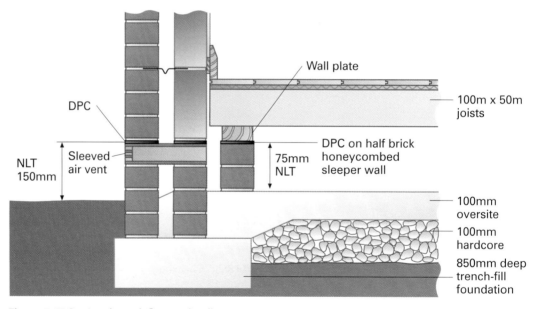

Figure 7.63 Section through floor and wall

The following pages describe the steps in fitting floor joists.

Step 1 Bed and level the wall plate onto the sleeper wall with the DPC under it.

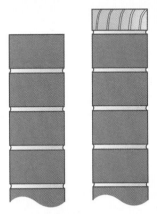

Figure 7.64 Bed in the wall plate

Step 3 Fix the first joist parallel to the wall with a gap of 50mm. Fix trimming and trimmer joists next to maintain the accuracy of the opening.

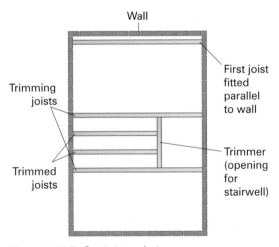

Figure 7.66 Fit first joist and trimmers

Step 2 Cut joists to length and seal the ends with preservative. Mark out the wall plate with the required centres, space the joists out and fix temporary battens near each end to hold the joists in position. Ends should be kept away from walls by approximately 12mm. It is important to ensure that the camber is turned upwards.

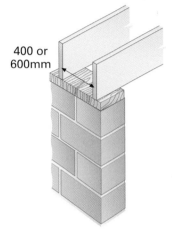

Figure 7.65 Space out joists

Step 4 Fix subsequent joists at the required spacing as far as the opposite wall. Spacing will depend on the size of joist and/or floor covering, but usually 400mm to 600mm centres are used.

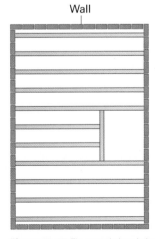

Figure 7.67 Fit remaining joists

Step 5 Fit folding wedges to keep the end joists parallel to the wall. Overtightening is to be avoided in case the wall is strained.

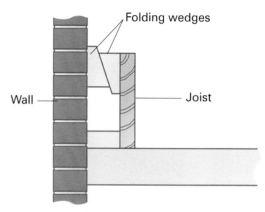

Figure 7.68 Fit folding wedges

Step 6 Check that the joists are level with a straight edge or line and, if necessary, pack with off-cuts of DPC.

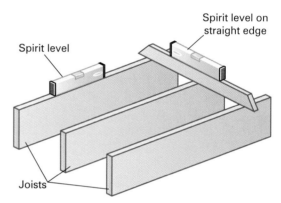

Figure 7.69 Ensure joists are level

Step 7 Fit restraining straps and, if the joists span more than 3.5m, fit strutting and bridging, described in more detail next.

Remember

It is very important to clean the underfloor area before fitting the flooring, as timber cuttings or shavings are likely to attract moisture

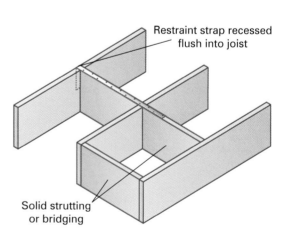

Figure 7.70 Fix restraining straps, struts and bridges

Strutting and bridging

When joists span more than 3.5m, a row of struts must be fixed midway between each joist. Strutting or bridging stiffens the floor in the same way that noggins stiffen timber stud partitions, preventing movement and twisting, which is useful when fitting flooring and ceiling covering. A number of methods are used, but the main ones are solid bridging, herringbone strutting and steel strutting.

Solid bridging

For solid bridging, timber struts the same depth as the joists are cut to fit tightly between each joist and **skew-nailed** in place. A disadvantage of solid bridging is that it tends to loosen when the joists shrink.

Solid bridging

Herringbone strutting

Here timber battens (usually 50 × 25mm) are cut to fit diagonally between the joists. A small saw cut is put into the ends of the battens before nailing to avoid the battens splitting. This will remain tight even after joist shrinkage. The following steps describe the fitting of timber herringbone strutting.

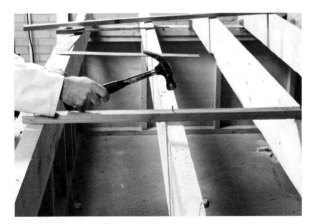

Space joists

Step 1 Nail a temporary batten near the line of strutting to keep the joists spaced at the correct centres.

Mark joist depths

Step 2 Mark the depth of a joist across the edge of the two joists.

Lay struts across two joists at a diagonal

Step 3 Lay the strut across two joists at a diagonal to the lines drawn in Step 2.

Cut to the mark

Step 4 Draw a pencil line underneath as shown in Step 3 and cut to the mark. This will provide the correct angle for nailing.

Fix the strut

Step 5 Fix the strut between the two joists.

Steel strutting

There are two types of galvanised steel herringbone struts available.

The first has angled lugs for fixing with the minimum 38mm round head wire nails.

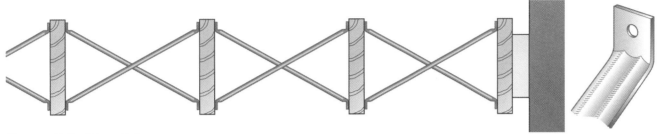

Figure 7.71 Catric® steel joist struts

The second has pointed ends, which bed themselves into joists when forced in at the bottom and pulled down at the top. Unlike other types of strutting, this type is best fixed from below.

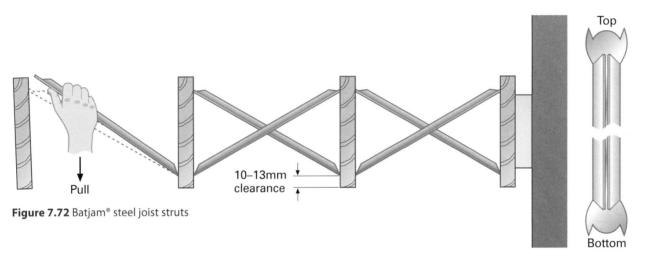

Figure 7.72 Batjam® steel joist struts

The disadvantage of steel strutting is that it only comes in set sizes, to fit centres of 400, 450 and 600mm. This is a disadvantage as there will always be a space in the construction of a floor that is smaller than the required centres.

Restraint straps

Anchoring straps, normally referred to as restraint straps, are needed to restrict any possible movement of the floor and walls due to wind pressure. They are made from galvanised steel, 5mm thick for horizontal restraints and 2.5mm for vertical restraints, 30mm wide and up to 1.2m in length. Holes are punched along the length to provide fixing points.

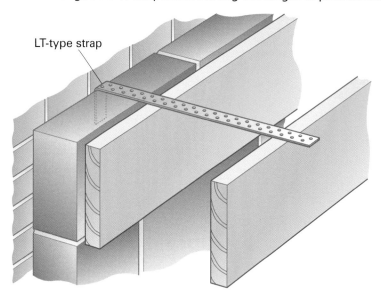

LT-type strap

Figure 7.73 Restraint straps for joists parallel or at right angles to a wall

When the joists run parallel to the walls, the straps will need to be housed into the joist to allow the strap to sit flush with the top of the joist, keeping the floor even. The anchors should be fixed at a maximum of 2m centre to centre. More information can be found in schedule 7 of the Building Regulations.

Floor coverings

Softwood flooring

Softwood flooring can be used at either ground or upper floor levels. It usually consists of 25 × 150mm tongued and grooved (T&G) boards. The tongue is slightly off centre to provide extra wear on the surface that will be walked upon.

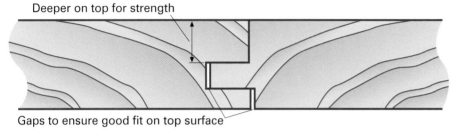

Figure 7.74 Section through softwood covering

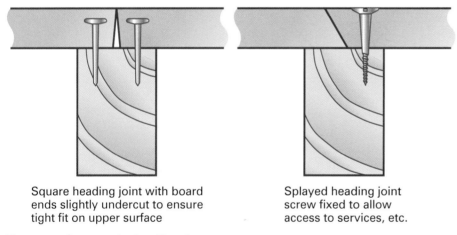

Square heading joint with board ends slightly undercut to ensure tight fit on upper surface

Splayed heading joint screw fixed to allow access to services, etc.

Figure 7.75 Square and splayed heading

When boards are joined together, the joints should be staggered evenly throughout the floor to give it strength. They should never be placed next to each other, as this prevents the joists from being tied together properly. The boards are either fixed with floor brads nailed through the surface and punched below flush, or secret nailed with lost head nails through the tongue. The nails used should be 2½ times the thickness of the floorboard.

The first board is nailed down about 10–12mm from the wall. The remaining boards can be fixed four to six boards at a time, leaving a 10–12mm gap around the perimeter to allow for expansion. This gap will eventually be covered by the skirting board.

There are two methods of clamping the boards before fixing:

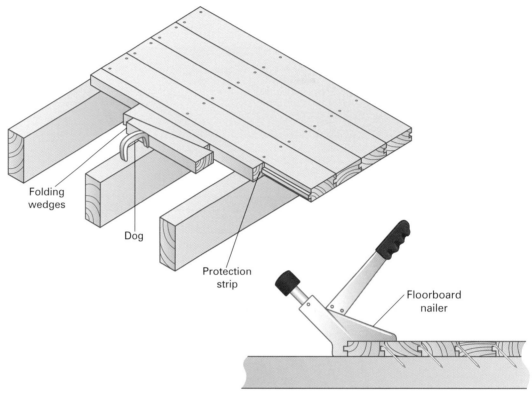

Folding wedges

Dog

Protection strip

Floorboard nailer

Figure 7.76 Clamping methods

Chipboard flooring

Flooring-grade chipboard is increasingly being used for domestic floors. It is available in sheets sizes of 2440 × 600 × 18mm and can be square edged or tongued and grooved on all edges, the latter being preferred. If square-edged chipboard is used it must be supported along every joint.

Tongued and grooved boards are laid end to end, at right angles to the joists. Cross-joints should be staggered and, as with softwood flooring, expansion gaps of 10–12mm left around the perimeter. The ends must be supported.

When setting out the floor joists, the spacing should be set to avoid any unnecessary wastage. The boards should be glued along all joints and fixed using either 50–65mm annular ring shank nails or 50–65mm screws. Access traps must be created in the flooring to allow access to services such as gas and water.

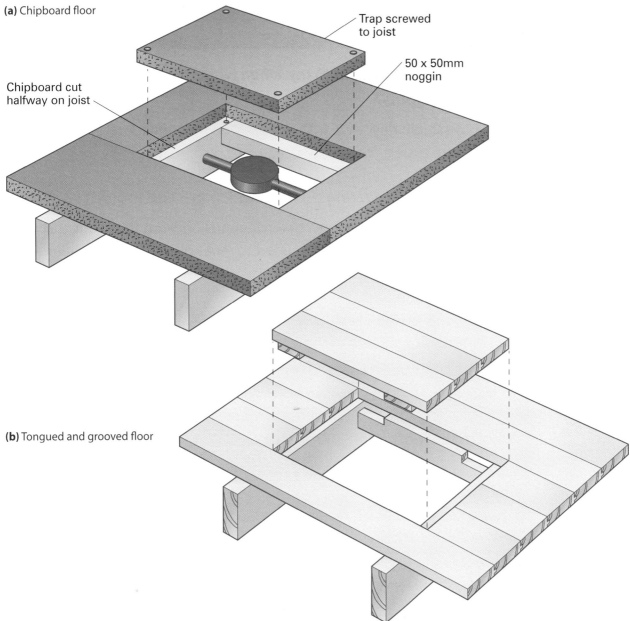

(a) Chipboard floor

Trap screwed to joist

50 x 50mm noggin

Chipboard cut halfway on joist

(b) Tongued and grooved floor

Figure 7.77 Access traps

On the job: Cutting in a garage roof

Chloe and Tyrone have been tasked with cutting in a garage roof with gable ends. They work out the angles and lengths, cut the pattern rafter, which they use to mark out all the others, then fit them. They are almost finished when the foreman turns up and tells them to stop and check it for level. Chloe and Tyrone put a level on the ridge board and find that it is way out of level. How could this have happened? What could have been done to prevent it?

FAQ

Which is the best method to use when working out roof bevels and lengths?

There is no single best way: it all depends on the individual and what they find the easiest method to use.

When it comes to covering the roof, which method should I use?

The type of roof covering to be used depends on what the client and architect want, and on what is needed to meet planning and Building Regulations.

Why does there have to be a 50mm gap between the rafters and the chimney?

Building Regulations state that there must be a 50mm gap to prevent the heat from the chimney combusting the timber.

Why are flat roofs only guaranteed for a certain amount of time?

Most things that you buy or have fitted have a guarantee for a certain amount of time and building work is no different. If the flat roof had a guarantee for 50 years, the builder would be responsible for any maintenance work on the roof free of charge for 50 years. Since the average life of a flat roof is 12–15 years, the builder will only offer a guarantee for 10 years.

Which type of flat roof decking is the best?

There is no specific best or worst but some materials are better than others. All the materials stated serve a purpose, but only if they are finished correctly, e.g. a chipboard-covered roof that is poorly felted will leak, as will a metal roof if the screw or bolt holes are not sealed correctly.

FAQ

How can I get the lead on a flat roof to fit into the brick/block work?

The lead is fitted into a channel or groove that is cut into the brick/block work by a bricklayer. The groove is cut using either a disc cutter or a hammer and cold chisel. Once the groove is cut the lead is fed into the groove, wedged and sealed with a suitable mastic or silicon.

I have laid chipboard flooring and the floor is squeaking. What causes this and how can I stop it?

The squeaking is caused by the floorboards rubbing against the nails – something that happens after a while as the nails eventually work themselves loose. The best way to prevent this is to put a few screws into the floor to prevent movement – but be cautious: there may be wires or pipes under the floor.

Knowledge check

1. Describe the difference between a single and double roof.

2. Explain the reason for having a mansard roof.

3. Explain the difference between a hip and a gable end.

4. What is a pitch line and where is it marked?

5. State a suitable way of fitting a wall plate.

6. State the distance the first and last common rafters must be away from the gable wall.

7. What is the dihedral angle (backing bevel) and what purpose does it serve?

8. What is the purpose of a roof ladder?

9. What type and size of nail should be used to fix bargeboard?

10. Why is it important that a carpenter learns Pythagoras' theorem?

11. Why do you not need to work out the plumb and seat cuts for the jack /cripple rafters?

12. State two different ways of forming a valley.

13. Why do you need to add extra rafters when trimming an opening?

14. State two different ways of providing a fall on a flat roof.

15. Name four different materials used when decking a flat roof?

16. What is bitumen?

17. What is the purpose of strutting a floor or flat roof?

18. What is the purpose of access hatches or traps in floors?

chapter 8

Maintaining components

OVERVIEW

In time, all components within a building will deteriorate. Timber will rot, brickwork crumble and even the ironmongery fail through wear and tear. These components need to be maintained or repaired to prevent the building falling into a state of disrepair. General building maintenance is now a recognised qualification, but traditionally this work is done by carpenters, as carpenters have the best links to other trades and the best understanding of what they do.

The maintenance of a building can range from periodically repainting the woodwork to replacing broken windows. This chapter will not cover simpler remedial tasks such as replacing a door handle, but will cover the following basic maintenance tasks:

- replacing broken glazing

- repairing and replacing mouldings

- repairing a door that is binding

- repairing damaged windowsills and doorframes

- replacing window sash cords

- repairing structural timbers

- minor repairs to plaster and brickwork.

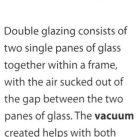
Replacing broken glazing

This is a job done largely by specialist glazing companies, but there is still a call for the carpenter to replace broken glazing. This section will look at replacing both single and double glazed windowpanes.

Single glazing

Single glazing is mainly found in older buildings such as factories or country houses. The glass is fixed using putty and held in place with either putty tooled to a smooth finish or timber beading.

The first thing to do when replacing a pane of glass is to remove the old pane. If the glass is smashed, you must remove the remaining pieces very carefully, wearing suitable gloves and other necessary PPE. If the glass is only cracked, it is safer to tape the glass to prevent it shattering, then tap it gently from the inside until it falls out.

Next you need to clean out the rebate, removing all the old putty so that the new pane of glass can be fitted properly. Use a sharp chisel, taking care not to damage the rebate.

Now fit the new pane of glass. There are several methods for doing this: use the method that matches the existing glazing in the building.

Method 1 Putty only

This is the traditional method, best done with a putty knife.

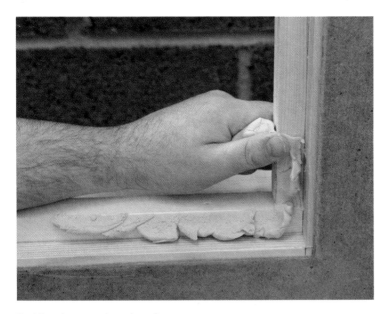

Pushing the putty into the rebate

First make the putty useable – when it is first taken out of the tub, putty is very sticky and oily, and must be kneaded like dough to make it workable. Then push the putty into the rebate, taking a small amount of putty in your hand and feeding it into the rebate using your forefinger and thumb.

Once the putty is all around the rebate, offer the glass into place starting at the bottom, then apply a little pressure to ensure that the glass is squashed up against the putty. Do not apply too much pressure, as this will break the glass.

Now insert glazing sprigs or panel pins to keep the glass in place until the putty sets. To avoid breaking the glass, place the edge of the hammer against the glass and slide it along the glass when driving in the pins or sprigs. If the hammer stays in contact with the glass, breakages are less likely. Drive the pins or sprigs in far enough that they are not in view, but not so far that they catch the edge of the glass and break it. Small panes need only one pin or sprig; larger panes will need more.

Take another ball of putty and feed it around the frame. Take the putty knife and, starting at the top, draw the knife down the window, squeezing the putty and shaping it at the same time. Repeat several times if necessary to get the desired finish. Remove any excess putty, including any on the inside that has been squeezed out from the rebate when the glass was pushed into place.

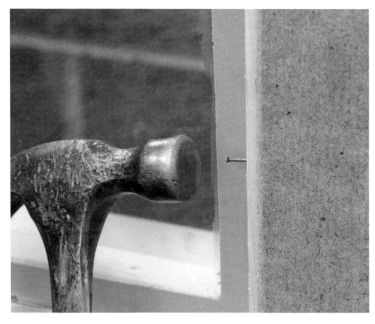

Sprigs being fitted

Putty being tooled

Did you know?

When using a putty knife, it can be useful to dip the knife into water to prevent the putty sticking to the blade and dragging the putty

Method 2 Putty and timber beading

This is similar to Method 1, following the same steps until the glass has been pushed into place. At this point, instead of the putty being tooled along the outside, a timber bead is used.

First place putty on the face of the bead that will be pushed against the glass, again using your forefinger and thumb.

Nailing the bead in place

Next fit the beads, which are usually secured using panel pins or small finish nails such as brads or lost heads. The beads should be fitted in sequence: top bead first, then the bottom and finally the sides.

Press the beads into place with enough force to squeeze the putty out, leaving just a small amount, then drive the pins or nails in as in Method 1. It can be difficult to put pressure on the bead and nail it at the same time, so ideally you need two people. Alternatively you can drive the nail into the bead a little way, but not far enough to penetrate the other side, then push the bead into place with one hand and drive the nail home with the other.

Once all the beads are fitted, remove the excess putty from both the inside and outside of the glass.

Method 3 Silicone and beads

This is identical to Method 2 but uses silicone instead of putty. It is often preferred as it is quicker and not as messy.

With advances in technology there are other ways of glazing. Rubber bead, tape or strips can be used to glaze single panes, but is more commonly used on double glazing. We will cover this method in the next section.

Double glazing

Double glazing can be found in either timber or UPVC windows. With UPVC double glazing, there are many different ways to install the glass using different types of beads and rubber components, so it is best to refer to the manufacturer when changing a unit. Here we will concentrate on replacing double-glazed units on timber windows.

When replacing double glazing, silicone and putty are rarely used, as a pre-made rubber bead, tape or strip is preferred.

Remember

With silicone, take care not to put too much on: once the glass or the timber bead is pushed into place, the silicone spreads more than putty but does not clean off glass as easily

Types of rubber bead, tape and strips

Various types of bead, tape and strip are available. Just a few examples are shown in Figure 8.1.

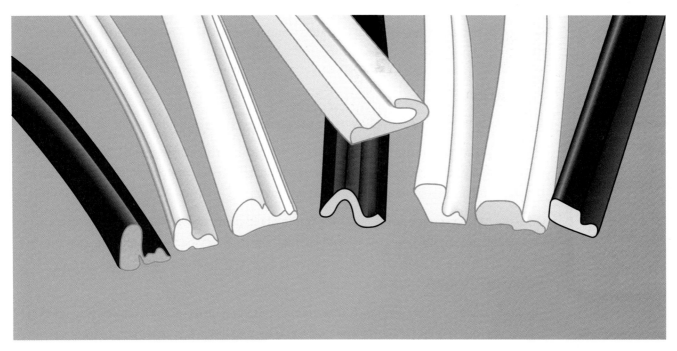

Figure 8.1 Some of the rubber beads, tape and strips available

First you need to remove the damaged or broken unit. As with single glazing, if the unit is just cracked, use tape to stop it shattering. Then remove the timber beads that keep the window in place, taking care not to damage them or the rubber if they are to be used again.

Now, wearing the correct protective clothing, carefully take out the glass. If the glass was installed with double-sided sticky rubber tape, run a sharp utility knife around the inside to free the glass. Next clean the beads and the rebate and, if you are using double-sided sticky rubber tape, apply this to the rebate.

Now the new double-glazed unit can be fitted and the beads pinned back into place.

Take care as some manufacturers have special insulated units that require the glass to go in a specific way. Such units carry a sticker that states something like *'this surface to face the outside'*.

Did you know?

Double-sided sticky rubber tape is a security device to prevent burglars removing the beads and taking the glass out

Repairing and replacing mouldings

Mouldings such as skirting or architrave rarely need repairing – usually only because of damp or through damage when moving furniture – and in most cases it is easier to replace them than repair them.

Architrave

To replace architrave, you simply remove the damaged piece and fit a new piece.

First check that the piece you are removing is not nailed to an existing piece. Then run a sharp utility knife down both sides of the architrave so that when it is removed it does not damage the surrounding decorations or remaining architrave.

Once the old piece is removed, clean the frame of old paint, give it a light sanding with sandpaper, then fit the new piece. Finally paint, stain or finish the new piece to match the rest.

Skirting board

Replacing skirting boards is slightly more difficult than replacing architrave. The way skirting boards are fitted could mean that the board you wish to replace is held in place by other skirting boards. Rather than remove the other skirting boards, you should cut or drill a series of holes in the middle of the board, splitting it in two so that you can remove it that way. Again, running a utility knife along the top of the skirting board will avoid unnecessary damage to surrounding decorations. Once the old skirting has been removed, the new piece can be fitted and finished to match the existing skirting.

You can replace other mouldings such as picture and dado rail in the same way as skirting boards, taking care not to damage the existing decorations.

Remember

When replacing mouldings, you may need to rub down the existing moulding work and re-apply a finish to both the old and the new to get a good match

Safety tip

Take care if gripper rods are fitted in front of the skirting, as this will make the replacing of the skirting more difficult and more dangerous

Repairing a door that is binding

One of the most common problems that occur with doors is that they **bind**. A door can bind at several different points, as you can see in the illustration below.

These binding problems are simple to fix.

- If the door is binding at the hinges, it is usually caused by either a screw head sticking out too far or by a screw that has been put in squint. In these cases, screw the screw in fully or replace it straight. If the hinge is bent, fit a new one.

- If the door is binding on the hanging stile, it may not have been back-bevelled when hung. In this case, take the door off, remove the hinges, plane the door with a back bevel and then re-hang it. Alternatively, it may be expansion or swelling due to changes in temperature that is causing the problem. In this case, take off the door, plane it and re-hang it.

Definition

Bind – a door 'binds' when it will not close properly and the door springs opens slightly when it is pushed closed

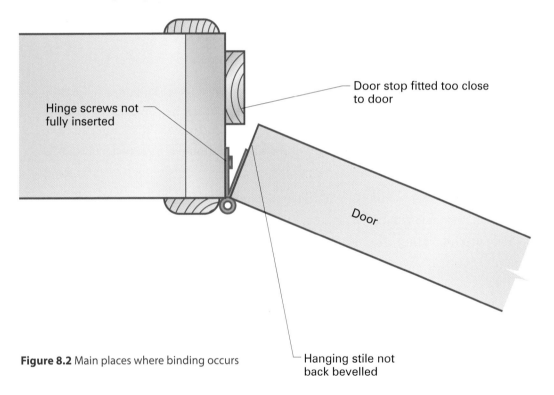

Door stop fitted too close to door

Hinge screws not fully inserted

Door

Figure 8.2 Main places where binding occurs

Hanging stile not back bevelled

- If the door is binding at the **stop on the hanging side**, the fitter may not have left a 1–2mm gap between the stop and the door to allow for paint, in which case the stop will have to be moved. If there is a doorframe rather than a door lining, there is no stop to remove: simply plane the rebate using a rebate plane.

Repairing damaged windowsills and doorframes

Provided they are painted or treated from time to time, exterior doorframes and windows should last a long time, but certain areas of frames and windows are more susceptible to damage than others. The base of a doorframe is where water can be absorbed into the frame; with windows, the sill is the most likely to suffer damage as water can sit on the sill and slowly penetrate it.

With damaged areas like these, the first thing to consider is whether to repair the damage or replace the component. With doorframes, replacement may be the best option but you must take into account the extra work required to make good: repairing the frame may cost less and require less work. With a rotten windowsill, replacement involves taking out the entire window so that the new sill can be fitted, so repair may be the better option.

The choice usually comes down to a balance of cost and longevity: as the experienced tradesperson, you must help the client choose the best course of action.

In this section we will look at repairing the base of a doorframe and a windowsill.

Repairing a doorframe

The base of the doorframe is most susceptible to rot as the end grain of the timber acts like a sponge, drawing water up into the timber (this is why you should always treat cut ends before fixing).

First check that the timber is rotten. Push a blunt instrument like a screwdriver into the timber; if it pushes into the timber, the rot is evident; if it does not, the frame is ok. Rot is also indicated by the paint or finish flaking off, or a musty smell.

If there is rot, repair it by using a splice, as follows.

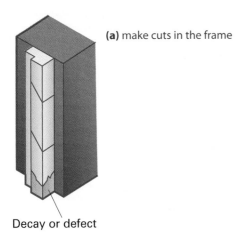

(a) make cuts in the frame

Decay or defect

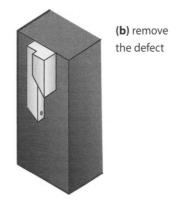

(b) remove the defect

Step 1 Make 45-degree cuts in the frame with the cuts sloping down outwards, to stop surface water running into the joints.

Step 2 Remove the defective piece and use a sharp chisel to chop away the waste, forming the scarf.

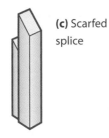

(c) Scarfed splice

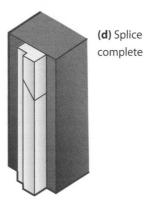

(d) Splice complete

Step 3 Use either a piece of similar stock or, if none is available, a square piece planed to the same size and shape as the original, making sure the ends are thoroughly treated before fitting to prevent a recurrence.

Step 4 Fix the splice in place, dress the joint using sharp planes and make good any plaster or render.

Figure 8.3 Stages in doorframe splicing

This method can also be used to repair doorframes damaged by having large furniture moved through them, but the higher up the damage on the frame, the more likely the frame will be need to be replaced rather than repaired.

Repairing a windowsill

As with a doorframe, this method uses a splice, as follows:

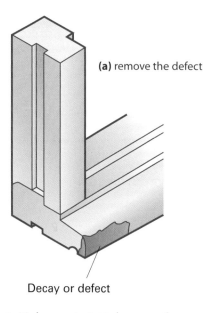

(a) remove the defect

Decay or defect

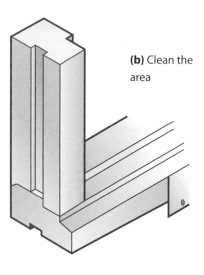

(b) Clean the area

Step 1 Make a cut at 45 degrees, then carefully saw along the front of the windowsill to remove the rotten area.

Step 2 Clean the removed area using a sharp chisel and then, as before, use either a piece of similar stock or, if none is available, a square piece planed to the same size and shape as the removed area.

Step 3 Give all cut areas a thorough treating with preservative, then attach the splice to match the existing windowsill shape and paint or finish to match the rest.

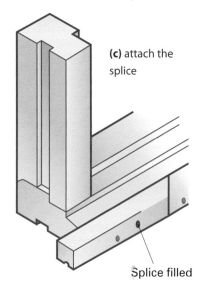

(c) attach the splice

Splice filled

Figure 8.4 Stages in windowsill splicing

Replacing window sash cords

As you saw in Chapter 5, box sash windows use weights attached to cords that run over a pulley system, to hold the sashes open and closed. The sash cord will eventually break through wear and tear, and will need to be replaced. This is not a large job so, if one cord needs attention, it is most cost-effective to replace them all at the same time.

There are various ways to replace sash cords. The following method is quick and easy and is done from the inside.

Step 1 Remove the staff bead, taking care not to damage the rest of the window. Cut the cords supporting the bottom sash, lowering the weight gently to avoid damaging the case. Take out the bottom sash, removing the old nails and bits of cord, and put to one side.

Step 2 Pull the top sash down and cut the cords carefully. Remove the parting bead and then the top sash, again removing the old nails and bits of cord, and put to one side.

Step 3 Remove the pockets and take out the weights, removing the cord from them and laying them at the appropriate side of the window.

Step 4 Slide the top sash into place at the bottom of the frame and, using chalk, mark the position of the sash cord onto the face of the pulley stiles (see distance A in Figure 8.5). Mark the bottom sash in the same way (distance B). Put the sashes to one side.

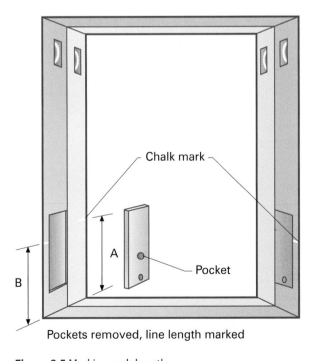

Pockets removed, line length marked

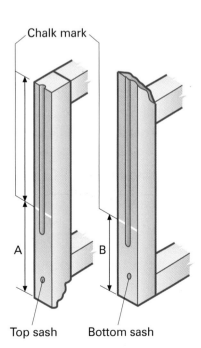

Top sash   Bottom sash

Figure 8.5 Marking sash lengths

Step 5 Next attach a **mouse** to the cord end, making sure it is long enough for the weighted end to reach the pocket before the sash cord reaches the pulley.

Step 6 Now cord the window. If only one cord is to be replaced, you can do this by feeding one pulley, then cutting the sash to length, re-attaching the end of the mouse to the sash cord and then feeding the next pulley. If you are replacing more than one cord, it is more efficient to feed the pulleys in succession and cut the cords after. In Figure 8.7 you can see one method for cording the window.

Mouse made of lead

75 mm

String about 2 m long

Cord

Figure 8.6 Mouse

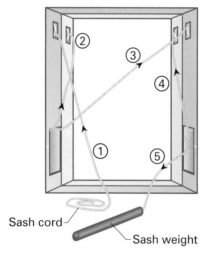

Sash cord

Sash weight

Figure 8.7 Cording a window

Feed the cord through the top left nearside pulley (1), out through the left-hand pocket, in through the top left far side pulley (2), then back through the same pocket. Then feed it through the top right nearside pulley (3), out through the right-hand pocket, in through the top right far side pulley (4) and back out of the right-hand pocket. Remove the mouse and attach the right-hand rear sash weight to the cord end (5).

Step 7 Working on the right-hand rear pulley, pull one weight through the pocket into the box and up until it is just short of the pulley. Lightly force a wedge into the pulley to prevent the weight from falling.

Pull the cord down the pulley stile and cut it to length, 50mm above the chalk line (when the window is closed and the weight falls back down to the bottom of the box, the weight will stop 50mm from the bottom of the frame as shown in Figure 8.9).

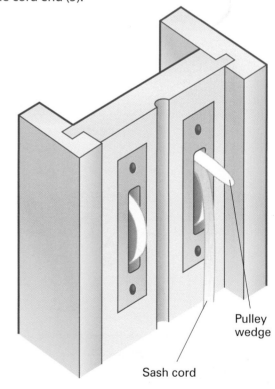

Pulley wedge

Sash cord

Figure 8.8 Wedged pulley

Step 8 Tie the other right-hand weight to the loose end of the cord. Still on the right-hand nearside, pull the cord again so that the weight is just short of the pulley and wedge it in place. Cut the cord to length, this time 30mm longer than the chalk line (when the window is closed the bottom sash weight will be 30mm away from the pulley, as shown in Figure 8.9).

Step 9 Fix the right-hand pocket back in place, then repeat Steps 7 and 8 for the left-hand side. Now all the pulleys are wedged and all the cords cut to length.

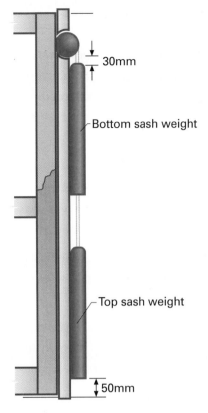

Figure 8.9 Sash weights at 50mm and 30mm

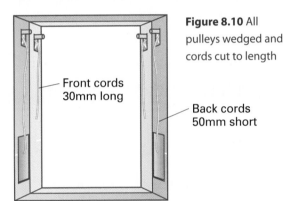

Figure 8.10 All pulleys wedged and cords cut to length

Front cords 30mm long

Back cords 50mm short

Step 10 Next fit the sashes, starting with the top sash. Fix the cord to the sash by either using a knot with a tack driven through it, or a series of tacks driven through the cord. Take care not to hamper the opening of the window, and do not use long nails as they will drive through the sash stile and damage the glass.

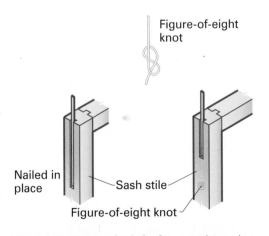

Figure-of-eight knot

Nailed in place

Sash stile

Figure-of-eight knot

Figure 8.11 Two methods for fixing cord to sashes

Step 11 Once the cord is fitted to both sides of the sash, slide the sash into place and carefully remove the wedges. Now test the top sash for movement and then re-fit the parting bead to keep the top sash in place. Now fit the bottom sash as in Step 10, test it and re-fit the staff beads, to secure the bottom sash in place.

Now test the whole window by sliding both sashes up and down. Finally, touch up any minor damage to the staff and parting beads.

Repairing structural timbers

The maintenance of structural timbers is vital, as they will almost certainly be carrying a load: joists carry the floors above, while rafters carry the weight of the roof. Because of this, structural repairs should be carried out by qualified specialists, so this section will give a brief understanding of the work that is involved.

Joists

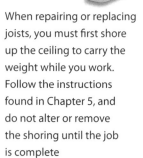

Safety tip

When repairing or replacing joists, you must first shore up the ceiling to carry the weight while you work. Follow the instructions found in Chapter 5, and do not alter or remove the shoring until the job is complete

Joist ends are susceptible to rot: they are close to the exterior walls and can be affected in areas like bathrooms if the floor gets soaked and does not dry out. Joists can also be attacked by wood-boring insects. For both rot and insect damage, the repair method is the same.

Shore the area, with the weight spread over the props, then lift the floorboards to see what the problem is (for the purposes of this example, we will use dry rot) and throw the old floorboards away.

Before making any repairs, you need to find and fix the cause of the problem – otherwise the same problems will keep on arising. Rot at the ends of joists is usually caused by poor ventilation: the cavity or the airbricks may be blocked, for example.

Next cut away the joists allowing 600mm into sound timber, and treat the cut ends with a suitable preservative.

New pre-treated timbers should be laid and bolted onto the existing joists with at least 1m overlap. New timbers should ideally be placed either side of the existing joist, and in place of the removed timber.

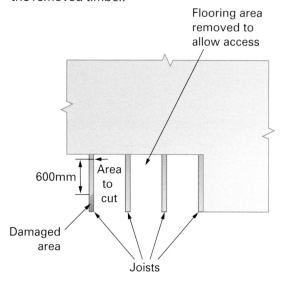

Figure 8.12 Cut away the rotten area of the joist

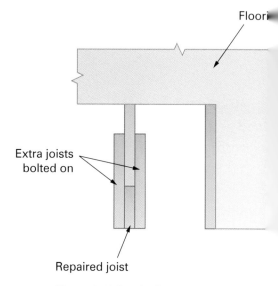

Figure 8.13 Repaired joist

Any new untreated timbers should be treated with a suitable preservative, before or after laying. Now slowly remove the props and make good as necessary.

Rafters

Rafters are susceptible to the same problems as joists, and especially to rot caused by a lack of ventilation in the roof space. Again the problem needs to be remedied before any repairs start.

Rafters are replaced or repaired in the same way as joists. They are easier to access, but shoring rafters can be difficult. In some cases it is best to strip the tiles and felt from the affected portion of the roof before starting.

For trussed roofs you should contact the manufacturer: trusses are stress graded to carry a certain weight, so attempting to modify them without expert advice could be disastrous.

Minor repairs to plaster and brickwork

During repair work there is always a chance that the interior plaster or brickwork may get damaged. Rather than call in a specialist most tradespeople will repair the damage themselves.

Repairing plasterwork

Plaster damage most often occurs when windows or doorframes are removed – the plaster cracks or comes loose – and is simple to fix with either ready-mixed plaster (ideal for such tasks) or traditional bagged plaster, which is cheaper.

Bagged plaster needs to be mixed with water and stirred until it is the right consistency. Some people prefer a thinner mix as it can be worked for longer, while others prefer a consistency more like thick custard as it can be easier to use.

Whichever type of plaster you use, the method of application is the same.

First brush the affected area with a PVA mix, watered down so it can be applied by brush. This acts as a bonding agent to help the plaster adhere to the wall.

Next use a trowel or float to force the plaster into the damaged area. If it is quite deep you may have to part-fill the area and leave it to set, then put a finish skim over it. Filling a deep cavity in one go may result in the plaster running, leaving a bad finish.

When the plaster is almost dry, dampen the surface with water and use a wet trowel to skim over the plastered area, leaving a smooth finish. Once the plaster is dry, the area can be redecorated.

Repairs to exterior render are done in the same way, using cement instead of plaster.

Remember

Attempt only minor repairs unless you are fully trained. Otherwise you could end up doing more harm than good!

Plaster being applied and trowelled

Repairing brickwork

With repairs to brickwork, one of the first problems is finding bricks that match the existing ones, especially in older buildings. After that, the method is as follows.

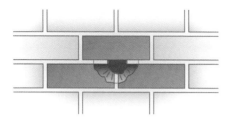

(a) Step 1

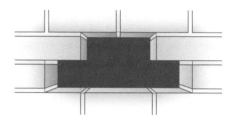

(b) Step 2

Step 1 Make a hole in the centre of the area, to allow the removal of the bricks. Bricks rarely come out whole, but care must be taken not to damage the surrounding bricks.

Step 2 Remove the bricks and carefully clean away the old mortar using a cold chisel.

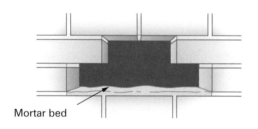

Mortar bed

(c) Step 3

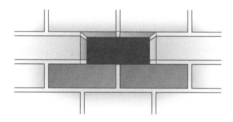

(d) Step 4

Step 3 Lay a mortar bed on the base of the opening.

Step 4 Place the first two bricks, making sure mortar is applied to all the joints.

Step 5 Place the last brick, again making sure there is mortar between all the joints, then use a pointing trowel to point in the new bricks to match the others.

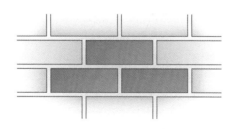

(e) Step 5

Figure 8.14 Stages in repairing brickwork

On the job:
Repairing a binding door

Kelly has been called to an office block to repair a door that is opening by itself. She checks the hinges, hanging stile and door stop for binding but they are all fine.

1. What could be causing the door to open?

2. How could this be fixed?

FAQ

Do I have to use putty or silicone to re-fix a pane of glass or can I use an adhesive?

Like other substances, glass expands and contracts slightly depending on the weather. This means that silicone or putty must be used as they remain flexible even when set. Adhesive sets rigid so, if any expansion or contraction occurs, the glass will break.

Can I put a double-glazed unit in a window or door that originally contained a single pane of glass?

Yes, as long as the rebate is wide enough to accept the unit and beading.

Can I do a scarf repair on skirting or architrave?

Yes, but it is often easier to replace mouldings than to repair them.

Should I go on a plastering course to help with repairs?

This is not necessary as the repairs you would be doing are not large, but if you feel that it would benefit you, yes.

Knowledge check

1. Why is double glazing preferred to single glazing?

2. When replacing a double-glazed unit why is it important that the glass is fitted in a specific way?

3. State the three methods used when single glazing and state which is best.

4. What is the purpose of double-sided sticky tape in relation to glazing?

5. Give a reason why mouldings may need to be replaced.

6. What is the term for cutting in a repair piece at 45 degrees?

7. Name the component attached to the end of a sash cord and used to help feed the cord through the pulleys.

8. Why should trussed rafters not be repaired by a non-specialist?

9. What is the main cause of rot in rafters?

10. What is the first step to take before applying plaster?

chapter 9

Woodworking machines

OVERVIEW

Woodworking machines are vital in producing components accurately and safely. The machining of timber has become more prevalent in recent years. Technology has allowed the building of machines that can transform stock timber into any size or shape, making the woodworker's job far easier. Every carpenter or joiner will at some stage come across a woodworking machine, whether it be a combination planer in a workshop or a table saw on site. It is important that you understand the principal uses, safety considerations and set-up of woodworking machines. This chapter will cover:

- general safety of woodworking machines
- table saws
- planers (surface, thickness and combination)
- band saws
- mortise machines.

General safety of woodworking machines

Compared to other industries, woodworking accounts for a large proportion of accidents. Woodworking machines often have high-speed cutters, and many cannot be fully enclosed owing to the nature of the work they do.

The use of woodworking machines was originally governed by the Woodworking Machine Regulations 1974. The introduction of the Provision and Use of Work Equipment Regulations (PUWER) 1992 superseded the 1974 Regulations, although regulations 13, 20 and 39 were still in use until the PUWER regulations were updated in 1998.

The PUWER regulations are explained in more detail in Chapter 1, but below are a few of the items relating to the safe use of all woodworking machines. Safety regulations relating specifically to a particular machine are noted in dedicated sections.

Safety appliances

Safety appliances such as push sticks/blocks and jigs must be designed so as to keep the operator's hands safe. More modern machines use power feed systems, eliminating the need for an operator to go near the cutting action. Power feed systems should be used wherever possible; in the absence of a power feed system, the appropriate push sticks/blocks must be used.

Working area

An unobstructed area is vital for the safe use of woodworking machines. The positioning of any machine must be carefully thought through to allow the machine to be used as intended. In a workshop environment, where there are several machines, the layout should be arranged so that the materials follow a logical path. Adequate access routes between machines must be kept clear, and there should also be a suitable storage area next to each machine to store materials safely without impeding the operator or others.

Floors

The floors around machines must be kept flat and in a good condition, and must be kept free from debris such as chippings, waste wood and sawdust. Any electricity supply, dust collection ductwork, etc. must be run above head height or set into the floor in such a way that does not create a trip hazard. Polished surfaces must be avoided and any spills must be mopped up immediately. Non-slip matting around a machine is preferred, but the edges must not present a trip hazard.

Lighting

All areas must be adequately lit, whether by natural or artificial lighting, to ensure that all machine set-up gauges and dials are visible. Lighting must be strong enough to ensure a good view of the machine and its operations, and lights must be positioned to avoid glare and without shining into the operator's eyes.

Heating

The temperature in a workshop should be neither too warm nor too cold, and the area should be heated if needed. A temperature of 16 degrees C is suitable for a workshop.

Controls

All machines should be fitted with a means of isolation from the electrical supply separate from the on/off buttons. The isolator should be positioned so that an operator can access it easily in an emergency. Ideally there should be a second cut-off switch accessible by others in case the operator is unable to reach the isolator. Machines must be fitted with an efficient starting/stopping mechanism, in easy reach of the operator. Machines must always be switched off when not in use and never left unattended until the cutter has come to a complete standstill.

Braking

All new machines must be fitted with an automatic braking system that ensures that the cutting tool stops within 10 seconds of the machine being switched off. Older machines are not required to have this, but PUWER states that all machines must be provided with controls that bring the machine to a controlled stop in a safe manner. The approved code of practice calls for employers to carry out a risk assessment to determine whether any machine requires a braking system to be fitted, and includes a list of machines for which braking will almost certainly be needed. If you are unsure, contact your local Health and Safety Executive.

Dust collection

Woodworking machines must be fitted with an efficient means of collecting the dust or chippings produced during the machining process.

Training

No person should use any woodworking machinery unless they have been suitably trained and deemed as competent in the use of that machine.

Maintenance

The machines should be maintained as per the manufacturer's instructions and should be checked over prior to use every day, with the inspection and any findings recorded in a log.

Table saws

Table saws – also known as rip saws, circular table saws and bench saws – come in a range of shapes and sizes.

Table saw

Flatting

The main functions of a table saw are flatting (cutting the timber to the required width) and deeping (cutting the timber to the required thickness).

With advances in technology and a new variety of saw blades available, the table saw is also capable of tasks ranging from rebating to creating housings. Here we will deal with the more traditional table saw, covering:

- table saw parts
- saw blades
- machine set-up
- safe operation.

Table saw parts

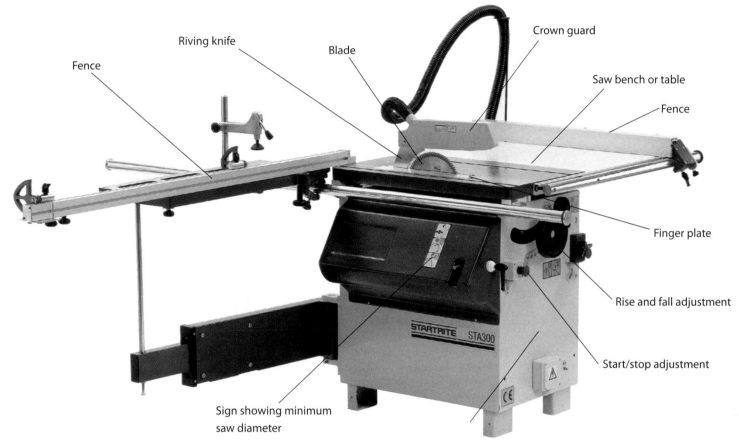

Parts of a traditional table saw

- **Saw bench or table** – the table or bench on which the saw is fitted. This should be extendable via rollers or extra out-feed tables to allow larger materials such as sheet timber to be machined

- **Blade** – the cutting tool

- **Crown guard** – a guard suspended over the top of the blade. It must be adjustable to ensure that as much of the blade is guarded as possible

- **Riving knife** – thicker than the blade, this acts as a spreader to prevent the cut from closing, which could cause binding

- **Fence** – a guide to give straight and accurate cutting, which should be accurately labelled and easily adjustable

- **Fingerplate** – a cover piece that sits over the spindle of the saw

- **Rise and fall adjustment** – a wheel used to adjust the height of the blade

- **Start/stop button** – this should be clearly labelled and within easy reach of the operator

- **Sign showing minimum saw blade diameter** – a legal requirement.

Saw blades

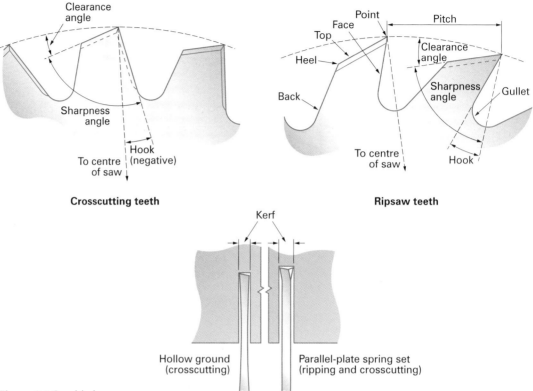

Figure 9.1 Saw blade

Several different types of saw blade are available, but here we will look at the standard types.

- **Pitch** – the distance between two teeth
- **Hook** – the angle at the front of the tooth. A positive hook is required for ripping, with an angle of 20–25 degrees for softwoods and 10–15 degrees for hardwood
- **Clearance angle** – ensures that the heel clears the timber when cutting
- **Top** – the angle across the top of the tooth
- **Set** – the amount each tooth is sprung out to give clearance and prevent the saw from binding
- **Gullet** – the space between two teeth, which carries away the sawdust
- **Kerf** – the width of the saw cut (not the width of the blade).

Tungsten carbide-tipped (**TCT**) blades are now preferred as they stay sharper for longer and do not have a set.

Calculating the rim or peripheral speed of a blade

The relation of the saw blade size to the speed of the spindle is crucial for the safe running of the saw (this is why the minimum size of the blade fitted must be shown clearly on the machine). A blade with a rim/peripheral speed of 100 metres per second can become stressed, causing a dangerous situation. A slow rim speed, on the other hand, means that the saw will struggle to cut the material.

To calculate the rim/peripheral speed, we must first know the diameter of the saw blade and the spindle speed. The diameter of the saw blade can be measured or found from the blade manufacturer; the spindle speed can be found from the saw manufacturer's handbook.

For this example we will use a saw blade with a diameter of 600mm and a spindle speed of 1600 revolutions per minute.

First work out the distance around the rim (the circumference of the blade), using the following calculation:

Circumference = π x diameter (where π is 3.142)

Circumference = 3.142×0.600

Circumference = 1.8852

This equals the distance travelled by 1 tooth in a single revolution of the blade.

Next find the distance travelled every minute by 1 tooth, by multiplying the circumference by the spindle speed:

$1.8852 \times 1600 = 3016.32$ metres per minute

Finally, to find the answer in metres per second, divide this answer by 60:

3016.32/60 = 50.272 metres per second.

A rim/peripheral speed of 50 metres per second is considered suitable.

If you are unsure about which blade size to fit, contact the manufacturer.

Machine set-up

The set-up of any table saw depends on the manufacturer and the type and model of saw, so be sure to check the manufacturer's handbook prior to using any machine. Whatever the make or model of your saw, there are certain safety measures you must take before use. You must ensure that:

- the machine is isolated from the power source prior to setting up
- all blades are in good condition, securely fixed and running freely
- all guards, guides, **jigs** and fences are set up correctly and securely
- suitable push sticks are available and at hand ready to use
- suitable PPE is available.

Definition

Jig – a guide or template attached to the machine, allowing the machining of timber to a variety of shapes

Safe operation

The operation of each table saw varies slightly depending on the task you are doing, but the basic operating principles remain the same.

First ensure that the checks have been done and the machine is set up safely. Wear the appropriate PPE and have a push stick at hand ready for use.

The machine can then be started and allowed to get up to speed before you offer the timber to the blade. Stand at the side of the piece being sawn and not directly behind it, as **kickback** may occur and can cause injury.

Using the fence/jig as a guide, slowly ease the timber into the blade using a steady amount of pressure: applying too much pressure will cause the blade to overheat, leaving burn marks on the face of the timber and even causing the blade to wobble, resulting in kickback. The right amount of pressure varies from material to material and between sizes of material but, generally, if you hear the saw straining or if there is a burning smell from scorching timber, you are applying too much pressure.

Keep the timber hard against the fence to ensure accuracy, and continue to feed it into the saw until there is no less than 300mm remaining to be cut. The last 300mm or more must be fed through using a push stick. The push stick should also always be used to remove the cut piece between the saw blade and the fence and any other off-cuts.

Definition

Kickback – where the timber being machined is thrown back towards the operator at speed, usually because the timber is binding against the blade

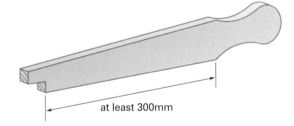

at least 300mm

Figure 9.2 Push stick

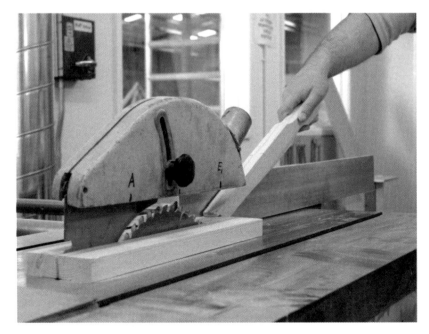

Correct use of a push stick

FAQ

What is binding?

Binding is where the timber being sawn closes or 'binds' over the blade so that contact with the face of the blade is made. This causes the timber to be thrown back at the operator ('a kickback').

What causes binding?

Binding can be caused by many different things including overexerting the blade by applying too much pressure and using a blunt or damaged blade. However, the most common cause of binding is high moisture content in the timber. Timber with a high moisture content has a tendency to move while being machined, which in some cases causes the kerf to close at the rear of the riving knife, applying pressure to the blade and causing binding.

How can I prevent the timber from binding when machining timber with high moisture content?

The best way is to have a second person who stands at the rear of the saw blade (well away from the blade) and drives timber wedges into the kerf to keep it from closing.

What is the best way to rip long lengths of timber?

Either have a long out-feed table or an extra worker at the rear to help guide the timber through the saw and support the weight of the timber as it passes through.

The timber that I am ripping down comes out a different measurement from the one I set the saw to. Why is this?

The gauge on the fence is out and must be recalibrated.

FAQ

The saw is not cutting straight. Why is this?

There are two main causes. Either the fence is not secured properly and is moving when you are feeding the timber through, or the timber is not being held tight against the fence throughout the operation.

What length does a push stick need to be and why?

A push stick must be between 300 and 450mm. Any less than 300mm and your hand will be too close to the operating blade; any more than 450mm and you are too far away to have proper control. Also, the longer the push stick the more chance there is of it springing out of place.

Planers

A hand-fed power planer is another essential tool in a woodwork shop as it can do in seconds what a craftsman would take hours to do by hand. This sort of planer is in essence the same as a portable power planer, but on a much larger scale and more accurate. All planers comprise an in-feed table, an out-feed table and a cutter block, into which either two or three cutting blades are housed.

Three main types of planer are available:

- surface planer
- thickness planer
- combination planer.

The combination planer is by far the best machine to have, as it can carry out all the operations of both surface and thickness planers. Given that the combination planer is essentially a combination of the other two, we will only look at the surface and thickness planers here.

Surface planer

As its name suggests, the surface planer creates a smooth and even surface on the piece of timber. It is most commonly used for two main operations:

1. **facing** – when the timber is planed flat and even on the widest side of the timber

2. **edging** – when the timber is placed flat and even on the narrowest side of the timber.

With the use of a fence the surface planer can also be set up to create rebates and splayed or angled pieces.

The surface planer works by passing the timber smoothly over an in-feed and out-feed table, in between which the cutter block is situated. The height of the in-feed table is adjustable to regulate the amount of timber removed in a single pass.

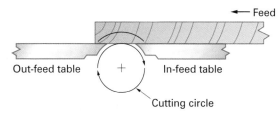

Figure 9.3 Cutter block, with out-feed and in-feed tables

The in-feed and out-feed tables are machined to be perfectly flat. When facing, provided the timber is held firmly to the surface of the out-feed table, a perfectly flat surface is produced. Edging is done by placing the faced side of the timber against the fence and running it over the machine.

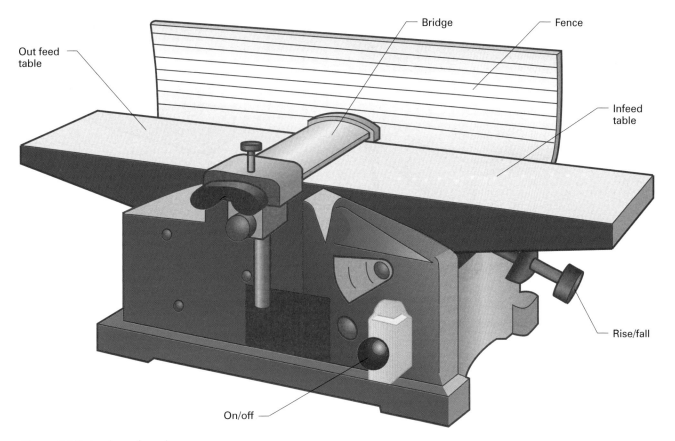

Figure 9.4 Parts of a surface planer

Figure 9.5 Telescopic bridge guard

The surface planer is fed mostly by hand, so great care must be taken to avoid your hands coming into contact with the cutter block.

The main way of protecting the user is through the use of suitable guards. The main guard used on a surface planer is the bridge guard. This is strong and rigid, and is usually made from aluminium so that if it comes into contact with the cutter block neither the guard nor block will disintegrate. Telescopic bridge guards are advisable as they can cover the full length of the cutter block; they must be wider than the cutter block.

Tunnel guards can also be used in conjunction with a push stick; push blocks should also be used when appropriate.

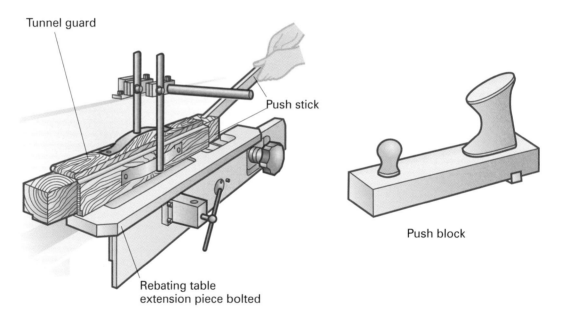

Figure 9.6 Tunnel guard and push block

There can be a tendency to feed timber over the cutter block too fast, which will leave a bad finish; to produce a good finish, it is best to go slowly.

Thickness planer

The thickness planer planes the timber to the required width or thickness, usually after the surface planer has faced and edged the timber. The thickness planer can also be used with a variety of jigs to create simple mouldings such as window beads.

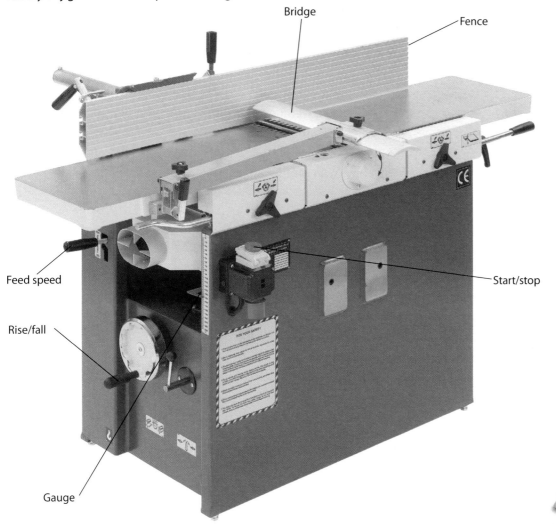

Bridge

Fence

Feed speed

Start/stop

Rise/fall

Gauge

Figure 9.7 Parts of a thickness planer

The thickness planer has a power feed system, usually in the form of four rollers. Two idle rollers are fitted at the bottom on the rise and fall table, and the two rollers above these are driven rollers, which propel the timber through the machine. The first upper roller is usually serrated so that it can grip the sawn timber better, while the second power roller is smooth so as to not damage the finished surface. To prevent kickback both upper rollers are spring-loaded, as are the two pressure bars. On most modern machines the serrated roller and front pressure bar are made in sections to allow more than one piece to be fed into the machine at the same time.

Safety tip

The thickness planer should carry a label stating whether more than one piece can be fed into the machine at one time. If you feed more than one piece into a thickness planer that doesn't have a sectional serrated roller, it may result in one of the pieces being thrown back at you at high speed

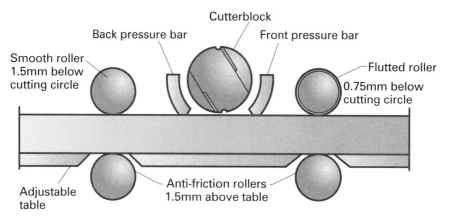

Figure 9.8 Section through a thickness planer

Operating the thickness planer is very simple: the timber is fed into the machine at one end and comes out planed to the required thickness at the other. You must be careful to ensure that the timber being fed in is of an appropriate thickness: too thick and planing the timber becomes very difficult; too thin and the timber will just pass through without being planed at all. The rise and fall table can be adjusted through a wheel or, on more modern machines, electronically via a button. A depth gauge should be in easy view so that the required depth can be set easily.

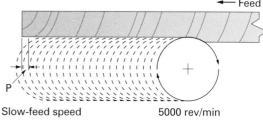

Figure 9.9 Pitch marks for slow- and fast-speed feeds

The thickness planer should have an adjustable roller-feed speed setting to help create a better finish. One of the most common problems with planing machines is pitch marks, where the surface of the timber is rippled. Pitch marks can be caused by having a fast in-feed speed or a slow cutter block speed.

Changing planer blades

Blades should be changed in line with the manufacturer's guidance. Special care must be taken to ensure that new blades are fitted correctly and in line with each other. Wrongly aligned blades may result in a poor finish: with one blade set above the other, only one blade is doing all the cutting, resulting in pitch marks. Poorly set blades may also pose a safety risk as they may shatter or chip.

FAQ

There is a small chip out of the blade on the surface planer I am using. Do I need to change the blades?

No. The fence adjustment should be moved so that you are planing in the area without the chip in it. Replacing blades every time there is a small chip can be an expensive practice.

How much material can I remove on a single pass with a surface planer?

It is not recommended to remove more than 3mm in a single pass.

I find it difficult to get a good finish when surface planing long lengths of timber.

When machining long lengths, it is best to have two people: one at the front and the other at the back, supporting the weight.

What PPE do I need to wear when operating planing machines?

You should wear boots, gloves, goggles, ear defenders and, depending on the timber being machined, a dust mask. You must also ensure that you have no loose clothing that could get caught in the cutter block.

The thickness planer I am using is creating pitch marks. How can I solve this?

Check the machine's feed speed and adjust it so that the machine feeds more slowly.

How do I know if I can feed more than one piece of timber into the thickness planer at the same time?

There should be a label on the machine stating that only one piece can be machined at any time. If there is no label, check the manufacturer's instructions. If you are in any doubt, feed just one piece at a time.

Band saws

The band saw is mainly used for curved or shaped work, but with the addition of fences a band saw can be used for ripping and crosscutting too. A band saw consists of an endless blade that runs around two wheels, with one wheel mounted above the other. The wheels are encased in the machine to protect the user, and an adjustable table sits between the wheels, which is where the cutting action occurs.

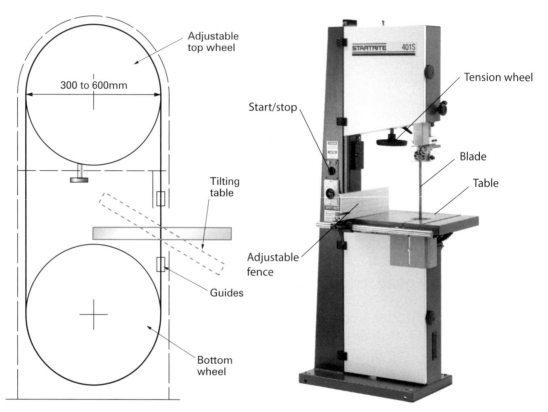

Figure 9.10 Wheels, table and blade for a band saw

Figure 9.11 Parts of a band saw

The bottom wheel of the band saw is driven directly by the motor, while the top wheel is driven by the blade. The top wheel is also adjustable to allow the blade tension to be set. Rubber tyres attached to the rims of the wheels help to stop the blade slipping, as well as preventing damage to the saw blade.

As well as running on the wheels, the blade is supported by saw guides situated above and below the saw table. The top guide is fitted with a guard to protect the user. The guides are fitted with thrust wheels, which prevent the saw blade being pushed back when the work is pressed against it. Blades come in a variety of sizes from 3mm upwards: the smallest blades are for more intricate, curved work; the larger blades are for ripping large, sectioned timber.

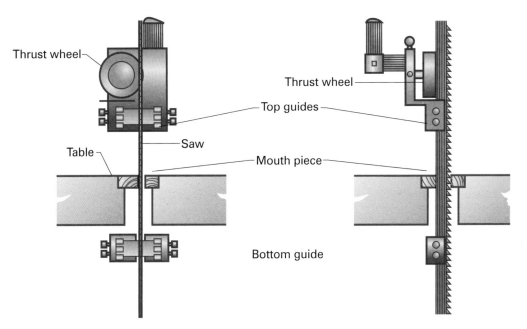

Figure 9.12 Guides

Remember

The exposed part of a band saw blade is guarded with an adjustable guard. It is vital that the guard is set prior to every use to leave as little of the blade showing as possible. The machine should already be set to the correct tension, but it is always better to check before use. A blade that is loose will come off the wheels and may snap; a blade that is too tight can snap too

Setting up the band saw

As with all machines, the set-up depends on the manufacturer. Not every machine can do all tasks, so it is always best to refer to the manufacturer's handbook when setting up the machine.

Before use, it is important to check the tracking. If the wheels are not set correctly, the blade could come off the wheels or tilt backwards, running against the guide and damaging or even breaking the blade.

Using the machine

The machine's operations will be listed in the handbook. Always ensure the machine is set up correctly for the operation you need prior to use. When working intricate curves into timber with a band saw your hands can get close to the blade, so it is vital to use push sticks and the correct fence attachments. If a band saw blade does snap, there can be a loud bang (and it can be quite scary), but all band saws must be fitted with a brake designed to stop the machine as quickly as possible, preventing the wheels from continuing to drive the broken blade.

Changing the blade

If a blade does break, or the machine needs to be set up for a different use, the blade will need to be replaced. Again, the manufacturer's handbook should be followed, but here is a basic guide to changing a blade:

Step 1 Ensure that the machine is switched off and isolated from the power supply before making any adjustments.

Step 2 Open the top and bottom guard doors and move aside any other obstructions such as the blade guards and table mouthpiece.

Step 3 Lower the top wheel to remove the tension, then remove the old/broken blade.

Step 4 Fit the new blade, ensuring that the teeth are pointing in the right direction (downward in most cases), then readjust the tension so that it is set correctly.

Step 5 Set the tracking so that the blade is running true and will not come off the wheels. Spin the wheels a few times to check that the tracking is correct.

Step 6 Reattach the guards, etc., reset the thrust wheels and guides and close the top and bottom guard doors.

Step 7 Switch the machine on and let it run for a little while to ensure that it is tracked and tensioned properly before attempting to cut any materials.

FAQ

How do I know what tension to set the blade to?

There should be a sign or panel on the machine that tells you what tension each type of blade should be set to. If not, contact the manufacturer.

There is a smell of burning rubber when the machine is in use. What is causing this?

This is happening because the tracking is out, making the blade cut into the tyres.

There is a thin blade in the machine and I want to rip large timber. Do I have to change the blade?

Yes, the thin blade will not cope with the task and will break.

Do I have to throw out my broken or blunt blades?

No, they can be repaired. However, band saw blades are not expensive so it is often best to buy new.

There are sparks appearing when I use the saw. What causes this?

Sparks can occur when the thrust wheel is not set correctly or when the tracking or tension is not correct. Switch off the machine, isolate it and check the thrust wheel, tracking and tension.

Mortise machines

The hollow-chisel mortise machine is designed predominantly to chisel out mortises for mortise and tenon joints, but it can also be used to chisel out mortises for any purpose.

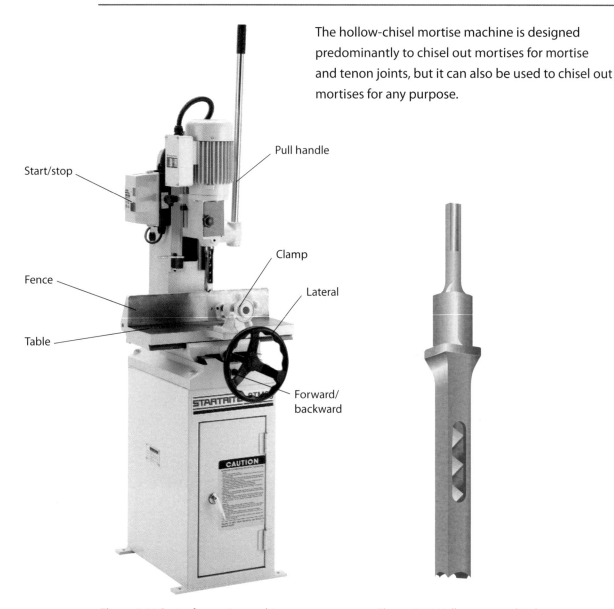

Start/stop

Pull handle

Fence

Clamp

Lateral

Table

Forward/
backward

Figure 9.13 Parts of a mortise machine

Figure 9.14 Hollow square chisel

A mortise machine consists of a revolving auger housed inside a hollow square chisel (this comes in various sizes, producing a range of mortises). The auger cuts most of the timber, while the chisel squares the hole up. The machine is operated much like a pillar drill, with a lever forcing the cutting action into the work piece. When setting the machine up it is important to ensure that there is a clearance of 2–3mm between the tip of the auger and the chisel: this prevents the tips of the auger and chisel overheating.

The work piece sits on a table and is clamped in place to prevent it from moving. The table can be moved forwards and backwards as well as from side to side; this means more than one hole can be drilled without removing the timber from the clamp.

Setting up and using the mortise machine

The mortise machine should be set up, and any tooling changed, as per the manufacturer's handbook. If the tooling has been changed, it is a good idea to run a test piece to ensure that the mortise is set up square; this prevents stepped mortising.

Once the machine has been set up correctly, the work piece can be clamped in place and any lateral or sideways adjustments can be made so that the chisel cuts in the correct place. When mortising full mortises, it is good practice to go only a little over halfway down the depth, then turn the timber over and complete the mortise from the other side; this avoids splitting out the bottom of the timber.

Safety tip

When machining timber of long lengths, make sure that they are adequately supported on any overhanging ends to prevent the work piece from lifting

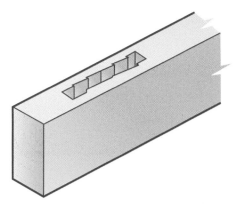

Figure 9.15 Stepping mortise caused by chisel not being set correctly

On the job 1: Feeding timber

Leroy and Judy are feeding timber into the thickness planer one piece at a time, as the machine is only a single feed machine. Leroy suggests that they feed more than one piece at a time so that they can be finished more quickly and go home. Judy says no, as she wants to do the job properly. What would you do? What could the outcome be of feeding more than one piece at a time?

On the job 2: Attracting attention

Simon is operating the table saw. He is wearing all the relevant PPE and has ear defenders on as it is very noisy. Matthew comes into the woodworking machine area to tell Simon that there is a phone call waiting for him. How should he get Simon's attention?

FAQ

How do I know if I have mortised halfway through?

Most machines come with a depth stop that can be set to any depth.

I want to mortise a 21mm mortise but the only chisels available are 18mm or 22mm. What do I do?

You cannot use the 22mm chisel, as it is too big. Instead, use the 18mm chisel and adjust the table laterally to complete the mortise.

The machine is creating stepped mortises. What can I do?

The chisel must be set up wrongly. You need to adjust it so that it is running square with the machine.

Knowledge check

1. What regulations govern the use of woodworking machines and what regulations did they supersede?

2. State a suitable temperature for a workshop.

3. Under new regulations, how long should it take a machine to come to a complete stop once the stop button has been pushed?

4. State the two main functions of a table saw.

5. What is the purpose of a crown guard?

6. What is the 'pitch' of a saw blade?

7. State the three main types of planer.

8. State three operations that can be carried out on a surface planer.

9. When planing what can cause pitch marks?

10. What are band saws mainly used for?

11. What is the purpose of the chisel on a mortise machine?

chapter 10

Mark, set out and manufacture joinery products

OVERVIEW

While the carpenter is generally employed on site carrying out first and second fixing and carcassing, the joiner is most likely to be found in a workshop manufacturing the components that the carpenter will fit on site. Whichever qualification you follow, you will need a good understanding of how basic components such as stairs are made.

This chapter will first cover general points about:

- setting out and marking.

Next, you will find details of the setting out, marking and manufacture of:

- windows
- doors
- stairs
- basic units.

Setting out and marking

The joints used in the manufacture of joinery components are covered fully in *Carpentry and Joinery NVQ and Technical Certificate Level 2.* This section will give a brief recap of the marking and setting out process.

Whatever you are setting out, it is best to start with a plan or a drawing. For joinery products, this is usually done to full scale, on a thin sheet of board (plywood, hardboard, etc.) better known as a setting out rod. The setting out rod is particularly crucial when manufacturing curved or complex work as it gives the joiner a true image of how the completed component will look.

Setting out rods should include horizontal and vertical sections through the components as well as elevations of any complex areas.

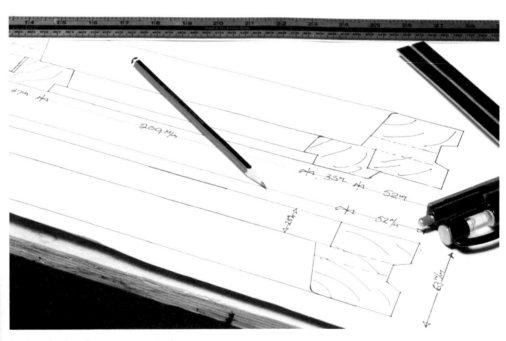

Rod marked up for casement window

Rods can be re-used by painting over them but it is a good idea to keep a record of the rods and store them for future reference.

Once the rods have been marked out, a cutting list can be made, the setting out rods can be stored to protect them from damage and the materials can be machined.

Now select the timber and cut to size, bearing in mind which will be the face and the edge. The face and edge are usually the two best adjacent sides, as they are most likely to be seen. Note the position and severity of any defects, as it may be possible to remove these when machining any rebates or grooves.

Next you must mark out the timber, which involves two main operations.

First you must apply face and edge marks, which serve as a reference point for the rest of the marking out and machining. The marks are shown right.

Now for the last stage in making out: the position of the joints, etc. To do this, you transfer the information from the setting out rods to the timber members. You should transfer the information as clearly as possible to avoid any confusion.

Now the joints can be cut and the component assembled.

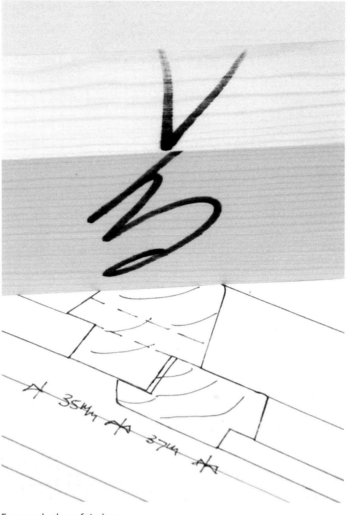

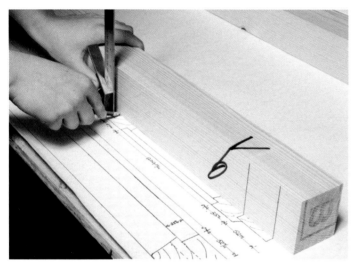

Transferring details from the setting out rod

Face and edge of timber

Windows

Windows are expected to stand up to the elements, so they usually need to be made from a solid construction.

In this section, our example will be a simple casement window, 900mm high and 500mm wide.

Setting and marking out a window

First you need to make setting out rods drawn full size to show the sections of the height and width.

Did you know?

The method of setting out and marking is virtually the same for every component. Remember this when you look at the following examples of manufactured components

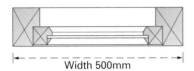

Figure10.1 Height and width rods

Use this drawing to produce the cutting list, then use the cutting list to prepare the materials for marking out.

When manufacturing a window, two different framing methods are used: one for the fixed parts and another for the moving parts. In our example, the fixed part is the window frame, and the moving part is the opening sash. For the frame, the horizontal members (head, sill) are mortised and the vertical members (jambs) are tenoned; for the moving sash, the opposite is true – the horizontal members are tenoned and the vertical members mortised.

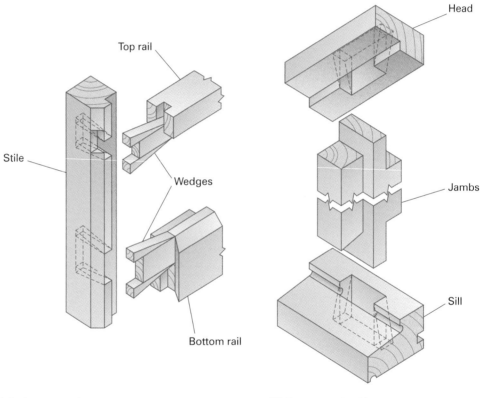

(a) Sash construction   **(b)** Frame construction

Figure10.2 Joints in frames and sashes

Remember

Don't forget that opposite members such as the head and sill or jambs should always be marked out in pairs

Once the materials are machined, you can mark them out by transferring the marks from the setting out rod to the various members.

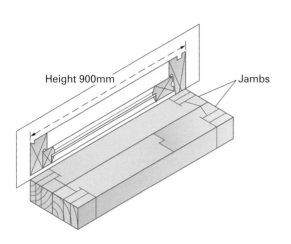

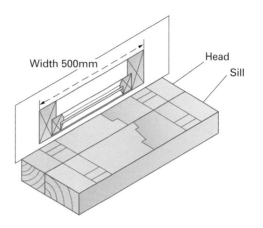

Figure10.3 Members marked out

Once all members are marked out, the joints, rebates, grooves, etc. can be machined ready for assembly.

Assembling a window

The first step in assembly is known as dry assembly. This is where the window is assembled without adhesive to check that the frame will be square and the joints will fit tightly. If the frame needs adjustment, this is the time to do it. After any adjustments, the frame can be glued up and clamped.

As soon as the frame is clamped but *before* the adhesive sets, you need to make two important checks.

First you must check for square, in one of three ways:

- **Use a large square** – simply try the square in all four corners to check for square.

- **Use the '3, 4, 5' method** – measure horizontally from one corner 300mm, then measure vertically down from the same corner 400mm: the distance between these two marks should be 500mm.

- **Measure the diagonals** – either simply measure from one corner to the other and check that the measurement is the same on the other diagonal, or use a **squaring rod**.

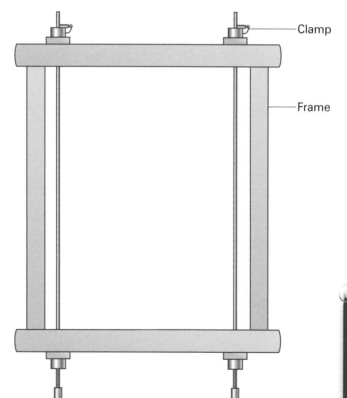

Figure10.4 Clamping up a frame

Definition

Winding – when a frame is twisted or skew

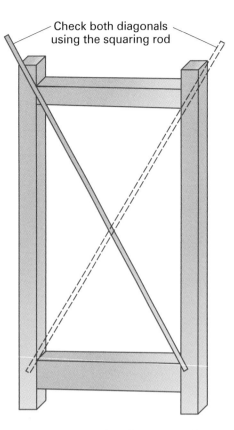

Check both diagonals using the squaring rod

Figure 10.5 Checking for square with a squaring rod

The second check you must make is for **winding**. The simplest way to do this is to ensure that the frame is laid on a flat surface and sight through the frame to check that there is no twist.

Now hammer wedges into the joints to tighten them up and use any fixings necessary. You can then fit any ironmongery, apply the finish and send the window out for fitting.

Doors

There are several types of door, but all doors come under one of two main categories.

- **Flush/hollow core door** – as the name implies, this is a hollow door with a frame around the outside, clad usually with hardboard or plywood. The interior of the door is packed with cardboard for strength, and a lock block is fitted to one of the stiles to allow a lock or latch to be fitted.

- **Framed door** – this is made from hardwood or softwood and constructed using either mortise and tenon or dowel joints. The frame is normally rebated to incorporate solid wood panels or glass, onto which beads will be fitted.

Flush doors and dowel-construction framed doors are normally factory-produced, so here we will look at what is involved in the production of a mortise and tenon framed door.

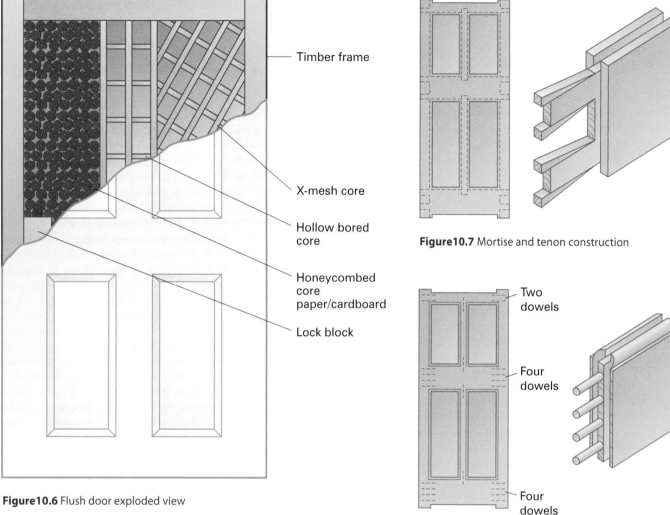

Timber frame

X-mesh core

Hollow bored core

Honeycombed core paper/cardboard

Lock block

Figure10.6 Flush door exploded view

Figure10.7 Mortise and tenon construction

Two dowels

Four dowels

Four dowels

Figure10.8 Dowel construction

Setting out and marking a door

As always, you start with setting out rods showing a section through the height and width.

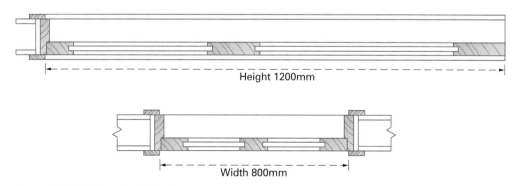

Height 1200mm

Width 800mm

Figure10.9 Height and width rods

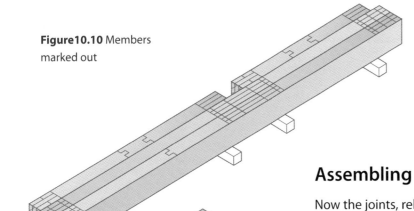

Figure10.10 Members marked out

You can use the rods to produce a cutting list, and use the cutting list to machine the members, remembering to apply the face and edge markings. Next transfer the marks from the rods to mark out all the members, remembering to mark out the horizontal members together and the vertical members together.

Assembling a door

Now the joints, rebates, etc. can be machined and the door can be finished. Remember to follow the correct sequence: dry fit, glue up, clamp up, check for square and winding, then clean up and apply finish. The door is now ready to be hung.

Stairs

Stairs are set out and marked out differently from doors and windows. To set out stairs, you first need all the dimensions, including the rise, going, etc.

Two templates are needed to mark out the tread and riser positions on the strings: the pitch board and the tread and riser template.

Remember

There are various regulations concerning stairs, and all dimensions must comply with these regulations. See Chapter 5 pages 79–80 for details

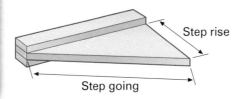

Step rise

Step going

Figure10.11 Pitch board

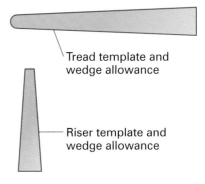

Tread template and wedge allowance

Riser template and wedge allowance

Figure10.12 Tread and riser template

Remember

Take time and care when making templates so that they are as accurate as possible

Setting out and marking stairs

First machine the strings to the required sizes and set them out on the bench, remembering to mark the face and edge marks.

Mark the pitch line using a margin template for accuracy.

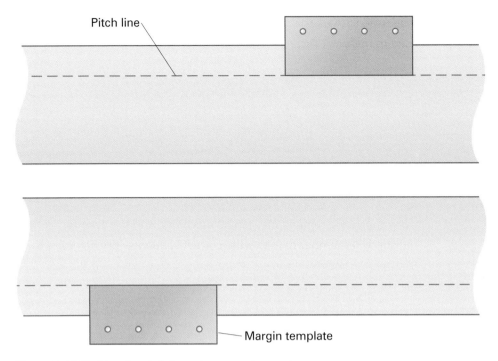

Figure10.13 Marking the pitch line using a margin template

Set a pair of dividers to the hypotenuse of the pitch board, then mark this distance all the way along both the strings: the two points of intersection will establish the tread and riser points relevant to the pitch line.

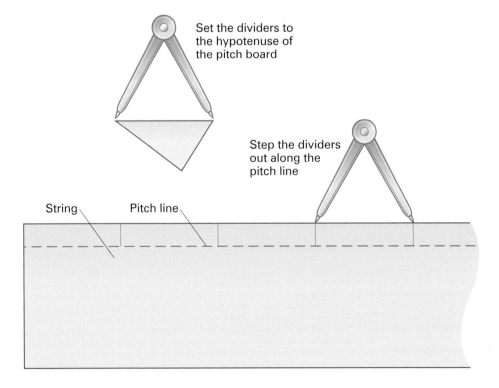

Figure10.14 Marking the tread and riser points

The pitch board can now be used to mark the rise and going onto the strings and the riser and tread templates can be used to mark out the actual position of each step.

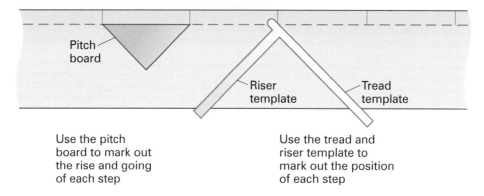

Use the pitch board to mark out the rise and going of each step

Use the tread and riser template to mark out the position of each step

Figure10.15 Marking the rise and going, and the position of the treads

Now router out the housings – it is best to use a stair housing jig combined with a router for this job.

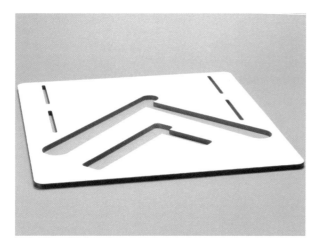

Router stair-housing jig

Stair housing jig used with router

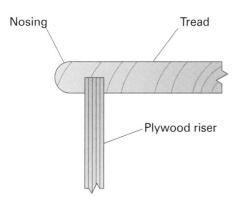

Figure10.16 Joint between tread and riser

Next the treads and risers can be machined. The treads usually have a curved nosing and a groove on the underside, to allow the risers to be housed into the treads.

The stairs can now be assembled.

Assembling stairs

Start by laying one string on the bench with the housings facing upwards. Put in place all the treads and risers, ensuring that the tread/riser ends and the housing are glued. Place the other string on top of the strings and risers, glue, then apply clamps.

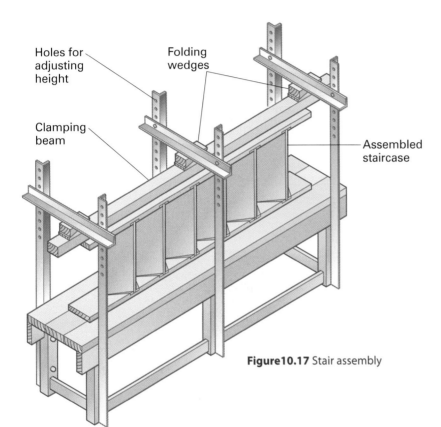

Holes for adjusting height

Folding wedges

Clamping beam

Assembled staircase

Figure 10.17 Stair assembly

Tread and riser assembly

Drive glued wedges into the housings to force the treads and riser to the front of the housing, making sure there are no gaps.

Once all the wedges are in place, check the front of the staircase to ensure that the treads and risers are situated perfectly. Now screw the bottom of the risers to the back edge of the treads. Finally fit glue blocks where the tread meets the riser, at the internal angle.

Once the adhesive has dried, remove the clamps, then clean up the staircase, protect it with hardboard or bubble wrap and send it on site to be fitted.

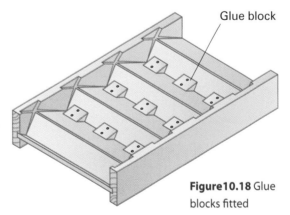

Glue block

Figure 10.18 Glue blocks fitted

Remember

If the string is not going to be seen or painted, it is good practice to screw through the strings into the treads making the stairs stronger

Basic units

The most common type of unit you will deal with is the kitchen unit: if you understand the principle of assembling these, you will be able to assemble any unit.

Most units are made from melamine-faced chipboard as this is light and easy to wipe clean, but other materials such as MDF, plywood and even solid timber can be used.

There are two main methods of unit construction.

- **Box construction** – also known as slab construction, this comprises vertical standards, rails and shelves, often with a plinth and bottom rail attached, with the remainder of the shelves sitting on adjustable brackets. With this method, the back panel is fixed to hold the frame square and make the unit sturdier.

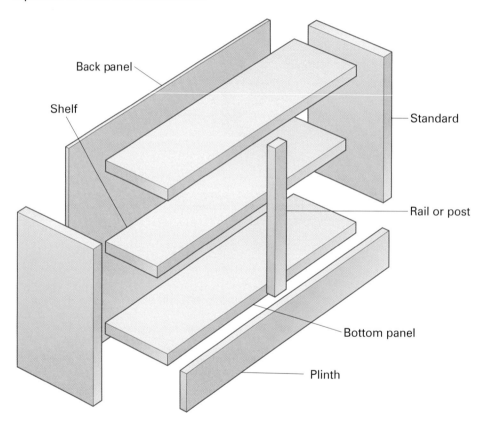

Figure10.19 Box construction

- **Framed construction** – also known as skeleton construction, this comprises either a pair of frames joined together with rails, or cross frames joined together by rails at the front and back.

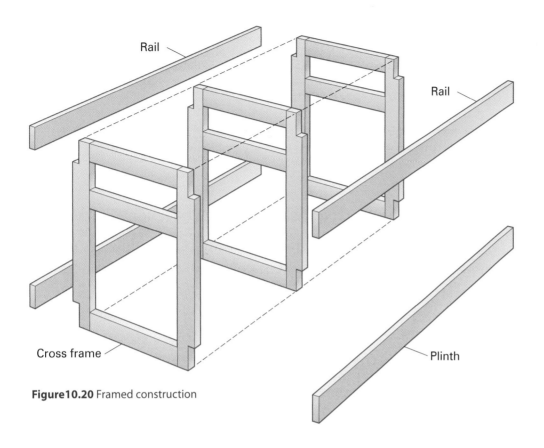

Figure10.20 Framed construction

Framed construction units usually have plinths built separately, and the frames are jointed using either mortise and tenon or halving joints.

Making these frames is very similar to making any frame. Most units can be mass-produced and bought cheaply, so there is less and less need for a joiner to construct units. Joiners are mainly used for specialist jobs where non-standard size units are required.

On the job: Producing setting out rods

Andy and Ravi have been tasked with producing a single casement window. Ravi says that he will go and produce the setting out rods if Andy gets some timber machined. Andy says that, since it is only one window, there is really no point in doing a setting out rod and it will be quicker to just mark it out as they go along.

What do you think? Who is correct? And why?

FAQ

Do face and edge marks have to be used?

No, they don't, but without them there is a good chance that mistakes will be made. Using face and edge marks is considered to be very good practice.

Which type of joint is best for a door: the mortise and tenon or dowel?

The mortise and tenon is the more traditional way, but the dowel is far easier. Both joints work well and the choice of which to use is down to the joiner or client.

Why are glue blocks fitted in stairs instead of just screwing down through the face of the tread?

Glue blocks prevent any movement once the stair is in use. They are preferred to screwing as screwing is unsightly and there is a risk of splitting the nosing.

Knowledge check

1. What drawings should setting out rods contain?

2. Why should rods be stored after the component is made?

3. What should the first stage in assembly be and why?

4. State two ways of checking a frame for square.

5. Name the two main categories of doors.

6. State the best way to cut out the housings on a stair string.

7. What is the purpose of glue wedges?

8. State four materials that units can be made from.

chapter 11

Mark, set out and manufacture complex joinery products

OVERVIEW

With the basics of component manufacture covered in the previous chapter, we will now look at more complex components. This chapter will cover the marking, setting out and manufacture of:

- louvre ventilator frames
- complex doors with shaped heads
- geometrical stairs
- laminated components.

Louvre ventilator frames

Louvre ventilator frames are commonly found in areas that require constant ventilation, such as boiler rooms. The ventilator is made up of a frame housing louvre boards, pitched at either 30, 45 or 60 degrees. Louvre frame ventilators are usually rectangular or square but for the purpose of this book we will look at shaped louvre ventilator frames. These are most commonly pitched, circular-headed or gothic.

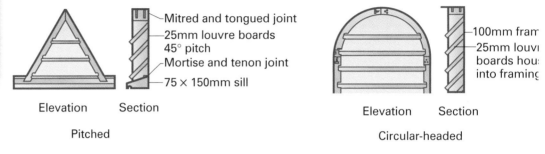

Elevation   Section

Mitred and tongued joint
25mm louvre boards
45° pitch
Mortise and tenon joint
75 × 150mm sill

Pitched

Elevation   Section

100mm fram
25mm louvi
boards hou
into framing

Circular-headed

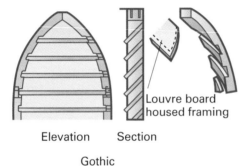

Elevation   Section

Louvre board housed framing

Gothic

Figure 11.1 Pitched, circular-headed and gothic louvre ventilators

The first style we will look at is the pitched ventilator.

Pitched ventilator frame

The first step, as always, is to draw out the frame full size. From this drawing we can get the true width and shape of the louvre boards.

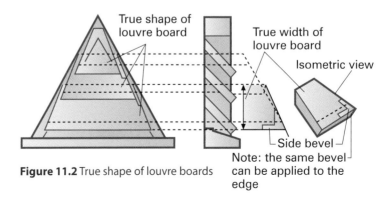

True shape of louvre board

True width of louvre board

Isometric view

Side bevel

Note: the same bevel can be applied to the edge

Figure 11.2 True shape of louvre boards

The louvres can then be marked out and cut, and the stock material for the frame can be machined.

The next stage is to mark out the frame joints and the housings, again using the full-scale drawing. The first step is to project the position of the housings from the section (A) onto the elevation (B). Machined stock (C) can be laid onto the drawing and the housings can be transferred from the elevation (B). The members should be marked out in pairs to ensure accuracy and that the finished frame is square.

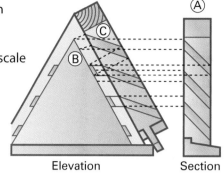

Elevation   Section

Figure 11.3 Marking out for housings

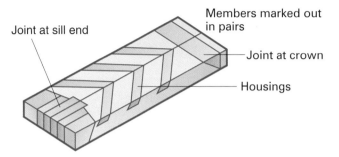

Figure 11.4 Marking out

The frame is now ready for assembly and finish.

Circular or curved ventilator frame

Circular or curved frames are more complex and the method of producing curved members is explained more fully in the next section.

The setting out process for circular headed and gothic frames is broadly the same as for the pitch frame, but the marking of the louvre positions is more difficult. It is best to use a square frame and a straight edge to mark the positions of the louvres on the face, and then use a pitched block to mark out the inner side of the frame.

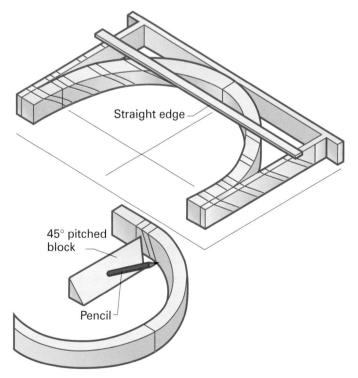

Figure 11.5 Marking for a circular head

Complex doors with shaped heads

Doors with shaped heads can provide a sense of grandeur but have to be made to measure, which can prove costly in today's market where most doors are mass-produced.

There are four main types of shaped head door:

- segmental
- semicircular
- gothic
- parabolic.

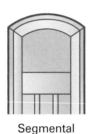   

| Segmental | Semicircular | Gothic | Parabolic |

Figure 11.6 The four main types of shaped head doors

When working with shaped head doors, many different types of joint can be used to create curved or arched members, for example, hammer headed keys, tenon joints or joints secured using handrail bolts.

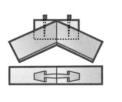 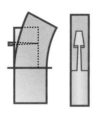 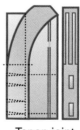

| Joint at crown | Joint at springing line | Tenon joint | Handrail bolt joint |

Figure 11.7 Hammer headed keys, tenon joints and handrail bolt joints

Basic method for producing a shaped head door

The example we use here is a semicircular door but whichever type of door is to be made, and whichever joints are used, the method is the same.

Working shaped members involves similar operations to working straight members, though greater care and skill are needed to maintain a uniform standard of craftsmanship. Vertical spindle moulders and routers are the best tools to use for creating the curved sections, and accurately made face templates are vital for cleaning and shaping the curved members. The height and width rods should be made as normal, but the circular head section must be marked out full size.

To mark out the shaped head, begin with the centre line and the springing line, where the joints at the stiles will be. Next mark on the stiles, and finally mark out the curved head with either a beam compass or radius rod.

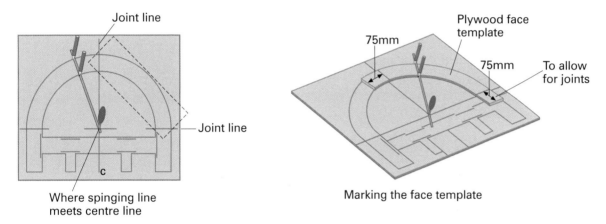

Figure 11.8 Setting out for a circular head

Figure 11.9 Face template

Next mark the face template from the drawing, making the template a minimum of 75mm longer at each end to allow for the joints to be constructed. The face template must be made from plywood at least 9mm thick.

Now machine the materials. The curved sections can either be formed on the vertical spindle moulder, with the face template used as a jig, or cut roughly on the band saw and tidied up using the face template and a router. Once the material has been planed and cut to the correct dimension, mark it out, remembering to mark out the stiles in pairs and the bottom, middle and frieze rail together.

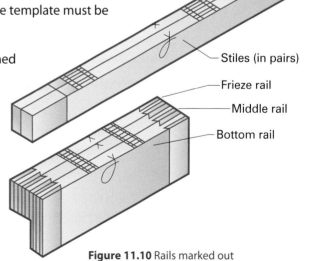

Figure 11.10 Rails marked out

227

The joints between the stiles and the rails are all mortise and tenon, so they can be manufactured and dry fitted to ensure a good tight fit. The joint between the stile and the circular head is a twin mortise and tenon – see Figure 11.11 for the detail.

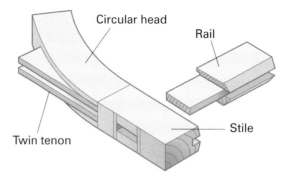

Figure 11.11 Twin tenon and rail

Joint the crown using handrail bolts with cross tongues for additional strength, as in Figure 11.12:

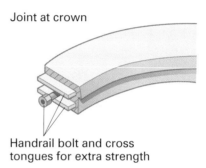

Figure 11.12 Crown joint

Once all the joints have been machined and tested for dry fitting, the door can be glued, assembled and clamped. When clamping the door, take care to ensure that the joints remain tight and the door does not go out of shape. Here is the ideal clamping method:

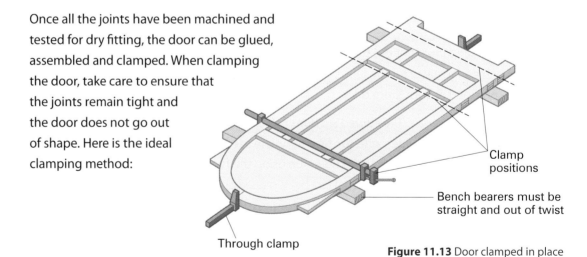

Figure 11.13 Door clamped in place

Once the adhesive is set, remove the clamps, and glaze and finish the door.

Geometrical stairs

Stairs are said to be geometrical when they have continuous strings and handrails. Geometric stairs are very complex, and building them requires a wide understanding of geometry. In this section you will find a brief overview of the geometry, and will see how to construct a basic geometrical stair.

Producing a geometrical stair

Our example here is a geometrical stair with a quarter over a **wreathed string** with a quarter-turn of winders (though this sounds complex, it is essentially a quarter-turn stair).

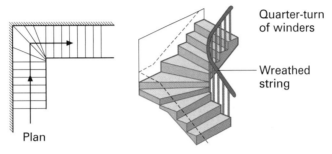

Plan

Quarter-turn of winders

Wreathed string

Figure 11.14 Quarter-turn stair

Definition

Wreathed string – continuous string that rises while turning

Notice that there is no newel post at the turn; instead the outer string is a continuous string, better known as a wreathed string. The wreathed string will be shown in more detail later but first we will concentrate on the wall strings.

The first thing, as always, is to do a drawing so that you can work out the size of the winders, etc. You must remember to keep within these regulations:

* the rise of the tapered steps must be the same as the rise of the other steps

* the tapered step must not be less than 50mm at the narrowest point

* the going of the tapered steps (measured at the centre of the steps) must be the same as the going of the other steps.

The stairs rise as they turn, so the wall strings need to be made wider. There are two ways to make the wall string: one is to make the string out of wider stock and cut away the waste, but this is very costly; the preferred, cheaper method is to attach pieces to the wall strings. The attached pieces should ideally be **biscuit-jointed** and glued for strength, though when the treads and risers are fixed to the string this will strengthen the joint.

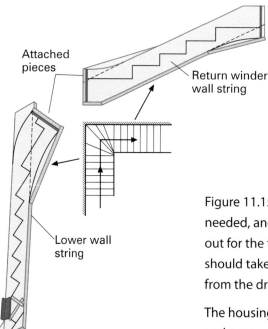

Figure 11.15 Wall strings

Figure 11.15 shows why the attached pieces are needed, and how the wall string should be marked out for the treads and risers. To maintain accuracy, you should take the sizes for marking out the winders, etc. from the drawings.

The housings for the risers and treads, and the tongue and grooved joint where the two wall strings meet, can all be machined.

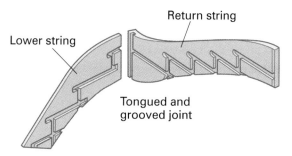

Figure 11.16 Joint between wall strings

Now to the wreathed string. Probably the most difficult aspect of making a wreathed string is finding the curved shape of the string – this is where a good grasp of geometry is required!

1. Start with a full-scale plan drawing of the stair (for this example we will use a 40mm string with a rise of 175mm and a going of 250mm).

2. Extend a horizontal straight line from point A the distance of the going (250mm) creating line A–B.

3. From point B draw a line down at 60 degrees to where it meets the outside of the string.

4. From point A draw a line down at 45 degrees to meet the 60-degree line, forming point C.

5. Draw a horizontal line from point C and a vertical line down from point A: where these two lines meet is point D.

6. Distance A–D will give the radius for a going of 250mm. From this the development of the string can be created.

The wreathed string is normally a cut string (as opposed to being routed) and consists of three parts. The wreathed part is the curved part, joined to a top and bottom cut string. This join can be done in three different ways, the most common of which is to use staves, which we will look at here (laminating is covered later in the chapter).

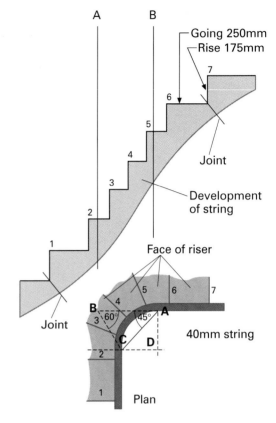

Figure 11.17 Geometry for wreathed string

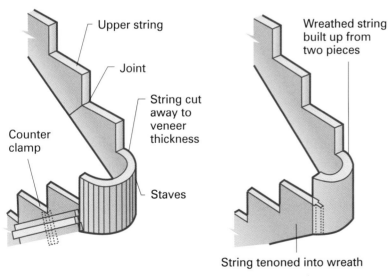

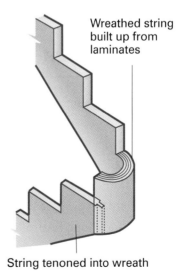

Figure 11.18 Wreathed string construction

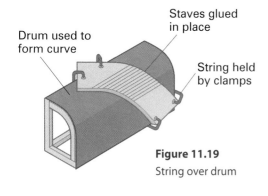

Figure 11.19
String over drum

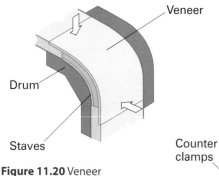

Figure 11.20 Veneer
being fitted over staves

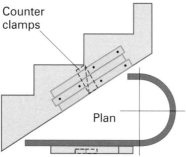

Figure 11.21 Counter clamps

The string starts out as a normal straight string piece. First you must reduce the string's thickness at the curved area to a veneer thickness of around 3mm. Once the section has been removed, place the string over a drum made to the shape of the curve and fix it in place, taking great care when bending the string. Glue a series of tapered staves into place and allow them to set.

If the string is going to be seen, and a quality finish is required, fit another veneer to the string to hide the staves.

Once the adhesive has dried, remove the string from the drum and joint it to the top and bottom cut string. The joint used depends on the thickness of the string and how the curve is formed. Three of these are familiar to you already – bridle, halving and mortise and tenon joints – so here we will look at the counter clamp method.

The counter clamp method is a simple but effective way of jointing the strings, working along similar lines to draw boring:

Once the wreathed strings are completed, fix the treads, winders and risers in place and the staircase is ready to be fitted and finished.

Laminated components

Laminated components are widespread throughout the construction industry, and are used more and more nowadays, the most basic form of laminate being plywood. Laminates are simply thin strips of timber that are glued together. In carcassing, particularly for roofing, laminates have an advantage over solid timber as laminated beams can be made to any length. Most laminated beams are now mass-produced and machined with specialist machines, but for the purpose of this book we will look at traditional methods for laminating a curved piece and a long beam.

Laminating a curved piece

Laminated curved pieces can be used for almost any purpose, and can be made to any radius, or even a variety of radii on a single piece to create a serpentine effect. Laminated timber can even be used for the shaped door heads discussed earlier in this chapter.

The amount of laminate needed depends on the thickness required, which in turn depends on the curve. For our example, the finished timber will be 60 x 60mm with a curve that is not too severe, so we will be using 3mm thick laminate.

Commonsense might say that $20 \times 3mm$ laminates will give you a 60mm finish, but this will probably not be the case as there will be a thin layer of adhesive between each laminate, increasing the finished size to more than 60mm. Nineteen laminates may be enough, but this may leave you slightly short of the 60mm. One option, depending on the precision of the finish required, might be to use 20 laminates, then plane the laminated component to the exact size using a spoke shave plane.

It is good practice to machine more laminates than you will use as this allows for any strips with shakes or dead knots, and any damage that might occur in the manufacturing process. The 3mm laminates should be sawn and ideally planed to the exact thickness. Ensure that all the laminates are uniform and clean, to create a good bond.

Once the timber is prepared to the correct size, it is ready for shaping. Shaping can be achieved by:

- **dry clamping** – the timber is placed into the jig and forced into shape by the clamp. This method yields poor results as the timber needs to be kept in the clamps for a long time, with no guarantee that it will keep its shape.

- **wet clamping** –the timber is soaked, then clamped in place. This produces better results than dry clamping, but again the timber must be kept clamped until it has dried out, which can be time-consuming. Also, with this method there is a good chance that the timber will spring back and not meet the required radius.

- **steam clamping** – the timber is placed into a steam box for a set time, then clamped into place. This is by far the best method: it does not take as long and gives the best results.

Timber is placed into the steam box and steam is pumped in. The heat allows the timber to be bent, while the moisture stops the timber becoming brittle and snapping. The length of time that the timber should be in the box depends on the type and thickness of the timber and the severity of the bend required, but 1 hour per 6mm thickness is a good guide time (it is better to oversteam the timber rather than understeam). Timbers should be placed into the box with piling sticks between them to allow the steam to circulate. After the required amount of time the timber should be removed.

Safety tip

When using a steam box, always wear the correct PPE as the timber will be hot and placed immediately into the jig and clamped in place

Did you know?

A steam box must be manufactured from a suitable material such as WBP plywood – otherwise the timber making up the box will break down from the effects of the steam!

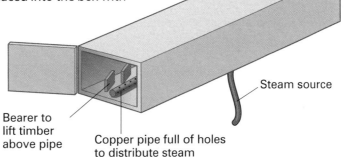

Bearer to lift timber above pipe

Copper pipe full of holes to distribute steam

Steam source

Figure 11.22 Steam box

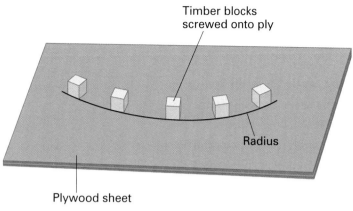

Timber blocks screwed onto ply

Radius

Plywood sheet

Figure 11.23 Typical jig

The clamping of the timber can be done in several ways, but the best way by far is to make up a jig.

This jig is an example only: any type of jig can be used.

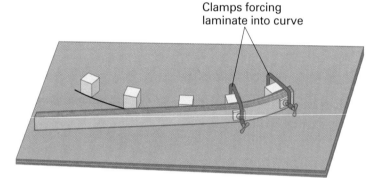

Clamps forcing laminate into curve

Figure 11.24 Timber clamped into a jig

With modern advances in adhesives it is now possible to glue up the timber straight out of the box and clamp it into the jig, but traditionally the timber is placed into the jig dry (i.e. not glued) and left overnight to dry out, then glued and clamped the next day.

Remember

When glueing and clamping a long beam, take care to ensure that the laminates are tightly clamped together and laid flat on bearers. This will produce a beam that is uniform and straight

Use the adhesive manufacturer's guidelines to decide the length of stay in the jig. Once dry, the laminated component can be cleaned up ready to be fitted. There may be a slight spring back once the clamps are released. This can be overcome by making the radius on the jig slightly more than required: when the spring back occurs it will leave the laminate at the correct radius.

Laminating a long beam

Laminating a long beam is a simple process with no need for steaming (unless the beam has a bend in it). A laminated beam can be made to any length because the laminates can be staggered.

The procedure starts with the required number of laminates planed to the required thickness, which for this example is 50 laminates at 5mm. The laminates are then glued together and clamped.

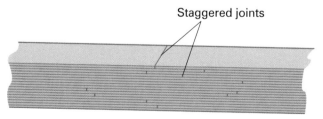

Staggered joints

Figure 11.25 Staggered laminates

On the job: Laminating

Derek and Janie are tasked with laminating a curved member to be used for the head of a doorframe. Neither of them has ever laminated anything before, but the supervisor tells them to cut up some laminates and stick them in the steam box for about an hour, set them up in the jig and then clamp them up. Has the supervisor given them enough information? What else would they need to know?

FAQ

How many louvres should there be in a ventilation frame?

The number of louvres depends on the client's wishes, but the spacing is very important, as they should not allow birds or other animals into the building.

Which is the best way to create a shaped head: using a band saw and router or using a spindle moulder?

Both ways are suitable, but using a spindle moulder creates the best finish and is quickest as it can be set to create moulded curved pieces in a single pass.

On a geometrical stair, does a wreathed string need to be a cut string?

No. The treads can be housed into the string, but with the curve this is very difficult so it is easier to have a cut string.

Do I need to cut veneers for both sides of the wreathed string?

This depends on the finish required. If the string is visible and has a varnish or stain finish, it is good practice to have veneers on both sides, but a painted string does not need veneers.

When laminating, what is spring back?

Spring back occurs when the clamps are released and the timber springs back into its natural state.

Knowledge check

1. State the three angles that louvres are normally pitched at.

2. Why are louvred ventilation frames necessary?

3. Name three types of shaped head doors.

4. State three jointing methods used when creating shaped head doors.

5. Why does a face template need to be accurate?

6. Stairs are said to be geometrical when they have what?

7. State one extra regulation that is required when making a staircase with winders.

8. Why are the staves that are fitted into the wreathed string tapered?

9. Give a reason why laminated beams are preferred to solid timber.

10. State a suitable guide time for steaming laminates.

chapter 12

Study skills/Advanced examinations

OVERVIEW

There are two main purposes of an Advanced Construction Award (ACA): to develop your practical skills and theoretical knowledge of your chosen trade, and to help you work towards an NVQ and advanced modern apprenticeship full framework.

There are two different types of examination for the Advanced Construction Award: a practical-based skill test called the **phase test**, and a full written theory test. The revision and the practice for the phase test have already been fully covered in the previous chapters of this book, so this chapter will concentrate on:

- what the examination will be about
- what types of question will be asked
- how to break down a question to find out what is really being asked
- how examinations are marked
- what preparation is required
- sample examination.

What the examination will be about

The ACA examination originally consisted of two separate papers: one that was trade-specific, and one that was a core paper covering areas such as health and safety, which was sat by all trades. A carpenter would sit the core paper and the carpentry paper, and a bricklayer would sit the core paper and the bricklaying paper. Each paper took between two and three hours, and both were full written rather than multiple choice.

In 2006 the ACA examination changed from two test papers to one. The examination now has two parts: part A dealing with the core, and part B dealing with the trade specifics. The question papers have an average of 30 questions – 10 for core and 20 for trade – and a time limit of between two and three hours. To achieve a pass, the candidate must do well in both parts A and B. If, for example, a candidate scores 19 out of 20 on part B but only 2 out of 10 on part A, they will fail the examination.

The grading system for the examination is as follows:

- **fail** – the candidate has not shown sufficient knowledge and has not passed the examination. The candidate will have to re-sit the examination

- **pass** – the candidate has shown sufficient knowledge to pass the examination

- **credit** – the candidate has shown very good knowledge and achieved above the pass mark

- **distinction** – the candidate has shown outstanding knowledge and achieved well above the pass mark.

What types of question will be asked

Three different types of question will be asked during your ACA examination:

- short answer questions

- structured questions

- questions which refer you to a handout or separate sheet.

Short answer questions

Short answer questions contain one or more problems, which you are required to answer with one or two words, a single sentence, a small sketch or a simple calculation. The length of the answer may vary from question to question depending on the topic and the question being asked.

Typical examples of short answer questions, with answers

> **Q: State the meaning of the following abbreviations:**
>
>   **a)   PUWER**
>
>   **b)   COSHH**
>
>   **c)   BSI**
>
> **A:**   a) Provision and Use of Working Equipment Regulations
>
>   b) Control of Substances Hazardous to Health
>
>   c) British Standards Institute
>
> **Q: Describe briefly why it is important to have regular health and safety meetings.**
>
> **A:**   To allow all staff and trades to find out what is happening and to have an input on the health and safety on the site.

Structured questions

Structured questions usually start with a statement containing some information, followed by a question or series of questions. The length of the answer required for each subsequent question will depend on what is being asked, as some of the questions may ask you to explain, state, list or even sketch your answer.

Typical examples of structured questions, with answers

> **Q: The Health and Safety (Safety Signs and Signals) Regulations apply to most work activities and premises.**
>
>   **a)   Briefly explain within your own occupation where a safety sign would need to be displayed.**
>
>   **b)   Sketch one safety sign and state what risk or hazard it is identifying.**

A: a) A safety sign would need to be displayed in a building site environment on the outside of the site telling all who plan to enter what hazards there are and what actions must be taken.

b)

Safety sign

This safety sign is stating that there is a noise hazard and it is informing the viewer that they must wear ear protection.

Q: **After climbing a ladder to reach the top of a scaffold you notice that a section of handrail is missing.**

   a) **State what three actions must be done to prevent injury.**
   b) **State who is responsible for checking the scaffold.**
   c) **Explain briefly what should have been done.**

A:

   a) The three things that must be done are: 1) inform all those around you and those working on the scaffold to leave the scaffold; 2) position a sign or tape of the ladder to prevent any one else accessing the scaffold; 3) inform the site agent and scaffolder.
   b) The site agent is responsible. Even though he or she may delegate responsibility to a ganger or working foreman, the site agent is still responsible.
   c) The scaffold should have been checked and, if it was found to be unsafe, then a red scaftag warning that the scaffold was unsafe should have been placed on the scaffold.

Remember

You must remember to write your name on the sheet and hand it in with your work or it will not get marked.

Questions that refer you to a handout or separate sheet

Some questions will refer you to a separate sheet contained within the question paper. This sheet will have a drawing or a table of sorts on it. The question will tell you what to do with the sheet, e.g. there may be a drawing of a site layout and the question will ask you to mark on the drawing a suitable place for a site office.

Typical example of a question that refers you to a separate sheet, with answers

Q: Complete the following accident report using an accident of your choice.

Report of an Accident, Dangerous Occurrence or Near Miss

Date of incident _____   **Time of incident** _____

Location of incident _____

Details of person involved in accident

Name _____   Date of birth _____

Address _____

_____

_____   Occupation _____

Date off work (if applicable) _____   **Date returning to work** _____

Nature of injury _____

Management of injury   ☐ First Aid only   ☐ Advised to see doctor

☐ Sent to casualty   ☐ Admitted to hospital

Account of accident, dangerous occurrence or near miss
(Continued on separate sheet if necessary)

[ ]

Witnesses to the incident
(Names, addresses and occupations)

[ ]

Was the injured person wearing PPE? If yes, what PPE? _____

_____

Signature of person completing form _____

Occupation _____   **Date** _____

A:

Report of an Accident, Dangerous Occurrence or Near Miss

Date of incident ___02/05/07___   Time of incident ___14:15___

Location of incident ___On site___

Details of person involved in accident

Name ___George Hill___   Date of birth ___12/07/75___

Address ___123 Any Street, Hometown HT5 3AS___

_____

_____   Occupation ___Carpenter___

Date off work (if applicable) _____   **Date returning to work** _____

Nature of injury ___Head injury___

Management of injury   ☐ First Aid only   ☐ Advised to see doctor

                                  ☑ Sent to casualty   ☐ Admitted to hospital

Account of accident, dangerous occurrence or near miss
(Continued on separate sheet if necessary)

> Falling tiles from ceiling roof hit the victim badly on the back of the head, knocking him unconscious. An ambulance was called and first aid given.

Witnesses to the incident
(Names, addresses and occupations)

> Peter Haining   56 House Lane
> Site supervisor   Hometown
>   HT1 5TQ

Was the injured person wearing PPE? If yes, what PPE? _____

___Safety helmet, safety gloves___

Signature of person completing form ___PHaining___

Occupation ___Site supervisor___   **Date** ___02/05/07___

How to break down a question

Certain questions, like the ones above, are pretty straightforward and easy to understand, but with some questions it can be harder to work out just what is being asked.

Example

Q: **Briefly explain, within the realms of your own occupational area, the purpose and reasons for the need of each of the following within your workplace.**

   a) **safety policy**

   b) **method statements**

   c) **risk assessments.**

Here is how we can break down the question.

In the first part – **Briefly explain, within the realms of your own occupational area, the purpose and reasons for the need of each of the following within your workplace** – there are two parts of the question that you can disregard: '**within the realms of your own occupational area**' and '**within your workplace**'. These simply mean that the examination concerns your work.

If we remove the unnecessary parts of the question the question reads:

Briefly explain the purpose and reasons why you need each of the following:

a) **safety policy**
b) **method statements**
c) **risk assessments.**

We can then break the three parts into three separate questions:

Briefly explain the purpose and reasons why you need a safety policy.

Briefly explain the purpose and reasons why you need a method statement.

Briefly explain the purpose and reasons why you need a risk assessment.

So what we have done is taken a complex question and broken it down to make it easier to understand and answer.

Briefly explain, within the realms of your own occupational area, the purpose and reasons for the need of each of the following within your workplace:

a) **safety policy**
b) **method statement**
c) **risk assessment.**

A:

a) A safety policy states what that company will do to promote and maintain a safe and healthy workplace. The reason for it is to ensure the company employees work to the policy: meaning they work within health and safety law.

b) A method statement states what will be done on a job and in what order as well as stating what safety precautions must be taken. The reason for this is to ensure a proactive approach is taken where everything is planned.

c) Risk assessments are produced prior to a task starting and they take into account all hazards and put in place plans to reduce any risks. The reasons for a risk assessment are similar to a method statement meaning a company will take a proactive approach to health and safety.

Some single sentence questions you may encounter ask you more than one thing.

Example

> **Q:** If you are heavily involved in conflict with other trades, how will this affect your work and how others treat you?
>
> Some people will understand that the question is asking you two things, but others may only pick up on one of the parts of the question. The question is really asking you:
>
> **If you are heavily involved in conflict with other trades how will this affect how others treat you?**
>
> AND
>
> **If you are heavily involved in conflict with other trades how will this affect your work?**

You can answer this type of question in two separate parts:

> **A:** If you have conflict with other trades they may end up disliking you and treating you badly.
>
> If you have conflict with other trades, then they may not help you, which could lead to you struggling to do your work.

Or you can combine the answers:

> **A:** If you have conflict with other trades, they may end up disliking you; this could result in them not helping you, which could lead to you struggling to do your own work.

Either way of answering this question is fine as long as it is answered fully.

Throughout the examination the questions will ask you to sketch, state, list, define, describe, explain, etc. Each of these key words should help you identify what type of question is being asked and what type of answer should be given.

Questions or parts of questions that contain the words 'list', 'state' or 'name' usually require you to give a short answer consisting of one or two words or a single sentence.

Example

> **Q: State three items of PPE.**
>
> **A:** 1. gloves, 2. boots, 3. goggles
>
> OR
>
> Three items of PPE are boots, gloves and goggles.
>
> Questions that ask you to define, describe or explain will require a longer answer.

Example

> **Q: Explain why we need PPE and when it should be used.**
>
> **A:** The reason we need PPE is to try and prevent accidents from occurring and PPE should be used when it is stated either by a sign or risk assessment. PPE should only be used as a last line of defence and should not be used as the sole way of preventing accidents.

Questions that contain the words 'sketch' or 'draw' will require you to provide drawings, remembering that a sketch need not be as detailed as a drawing. You must also look out for questions that contain two indicating words such as 'explain with the aid of sketches'.

How examinations are marked

The examination questions are set by a team of experienced people from industry and the awarding body. The examiners will have specific guidelines to follow when marking, and will have a selection of their marking sampled to ensure that they are not being too soft or too harsh.

Each question in the examination will have a specific maximum amount of marks.

Example

> **Q: Produce a safety checklist containing six items, for the erection of a step ladder.**
> (6 marks)

This question is worth six marks, so the examiner will be looking for six things in your answer.

Examiners can only mark what is put in front of them, and can only give marks for correct or relevant information. They cannot deduct marks for wrong information, poor spelling or grammar, although the examiner must be able to read your answer.

Marks are awarded for each question on its own merits. Examiners must not let a poor or good answer to a previous question reflect on the marking of the next question, ensuring that each question gets the marks it deserves.

What preparation is required

Examination papers are set to check that you have the knowledge required to back up your practical ability – they are not set to trip you up or trick you. It is not the examiners' fault if you fail to understand or misread a question and give a wrong or incomplete answer.

There are two main reasons why people fail examinations:

- lack of knowledge or understanding
- failure to prepare.

Lack of knowledge or understanding

You should only enter or be entered for an examination if you feel that you have sufficient knowledge and understanding to pass. Even if you do have good knowledge, if you do not revise for the exam you will reduce your chances of passing. Revision is an extremely important part of study and unless facts and information are repeated, constantly used or revised they will soon be forgotten.

Revision should ideally start at the beginning of a course and not at the end, just because there is an exam coming up. How you do your revision is an individual thing and not all people will be able to revise in the same way, but some of these methods may help.

- Rewrite rough or scribbled notes neatly at the end of lessons.
- Try to pick out, underline or highlight key words in your notes or textbooks: remembering these key words can help to remember a whole topic.
- Read textbooks or technical brochures.
- Answer the *Knowledge Check* section at the back of each chapter in this book.
- Get a colleague to ask you questions and if you give an unsatisfactory answer, make a note of the question and revise that topic.
- Do the practice examinations at the end of this chapter.
- Use any technique you have found useful previously.

Failure to prepare

The saying goes 'if you fail to prepare, you must prepare to fail' and examinations are no exception. Preparation for the examination is of paramount importance and there are certain things you must do.

Prior to the exam:

- Ensure you have revised fully.

- Arrive at the examination centre in plenty of time: an hour early is better than ten minutes late.

- Ensure you have spare pencils, pens, etc.

- Check if you need to bring anything with you such as calculators, etc.

- Ensure you have had a good meal beforehand: hunger pains will not help your concentration.

- Make sure you have been to the toilet: trying to hold it in or asking to go to the toilet will disrupt your concentration.

- Get the contact details for the examination room in case any one needs you in an emergency.

- Switch off your mobile phone.

Once you are in the examination room:

- Listen carefully to any instructions.

- Read the instructions at the top of the paper.

- Read through the whole paper, underlining any key words or points that may help you to answer that question later.

- Try to answer the questions you find easiest first as this will allow you to gain confidence and get some answers done (as opposed to sitting looking at a difficult question for an hour).

- Do *not* miss out or leave any question blank. If you are running short of time try to put something down that will at least get you some marks.

- If you have time at the end, re-read the questions and your answers to see if there is anything else you can add.

- Use the full time of the examination if needed. People have failed simply because they thought they had finished and left as soon as they could without checking their work.

- If you have to use additional sheets, ensure that your name and details are on them and that they are handed in with your work.

- Ensure that you have answered the question fully. If the question is worth three marks, make sure you have three marks' worth written down.

- If a question asks for five things and you can think of six then put six down, as one of the six you are thinking of may be wrong.

- Make sure your writing and sketches are legible: if you are struggling to read it, so will the examiners.

Once you have finished:

- Double and triple check your answers.

- Follow the procedure stated before the exam started to get an examiner's attention.

- If others are still working, leave the room and building quietly.

If you have revised and prepared well, then you will stand a good chance of passing.

Sample examination

When you are ready, try the following practice examination paper. It is best to practise using examination conditions, so try to ensure that you give yourself time to do the exam in one sitting.

Practice paper

You will need the following for this examination:

- answer book or blank paper with which to write your answers

- drawing instruments

- blue/black pen

- non-programmable calculator.

This exam will consist of two sections:

- Section 1: ten questions relating to core knowledge

- Section 2: twenty questions relating to the trade.

Candidates must achieve at least a pass mark on each section. The maximum marks for each question are shown in brackets.

Section 1 – Answer all **ten** questions. All questions carry equal marks.

1   Before commencing any work within an operational area, describe **four** important considerations. (4 marks)

2   State **four** reasons why the correct storage of materials is important. (4 marks)

3   With the aid of sketches, explain how on a mobile tower scaffolding the

   a) base wheels can be prevented from turning (2 marks)

   b) tower's stability can be increased allowing it to be built higher. (2 marks)

4   Describe briefly how the internal working environment can be improved during the construction of new buildings during the cold winter months. (4 marks)

5   Describe **four** features of a good transport route into and within a
    construction site.   (4 marks)

6   State **four** responsibilities of an employer under the Health and Safety at
    Work Act.   (4 marks)

7   State **two** important factors for

    a) holding regular safety meetings   (2 marks)

    b) holding regular progress meetings.   (2 marks)

8   State what initial actions are to be taken when there is a suspected theft
    on site.   (4 marks)

9   List **one** construction material that you are familiar with, which may cause problems for
    an operative when manual handling and state what precautions should be taken.
                                                                          (4 marks)

10 Describe **two** planning strategies used in your occupational area for preventing
    disruption of work due to

    a) non-delivery of materials   (2 marks)

    b) conflicting trade interests.   (2 marks)

Section 2 – Answer all **twenty** questions.

1   Describe with the aid of sketches, **two** ways of supporting a joist when
    abutting a brick wall.   (4 marks)

2   List **one** portable power tool that you are familiar with and state **four**
    precautions to take when using it.   (4 marks)

3   Calculate the distance one tooth travels in one minute on a circular saw blade
    if its diameter is 600mm and the spindle speed is 1700 rpm.   (2 marks)

4   Describe with the aid of sketches **two** ways to obtain a fall on a flat roof.   (4 marks)

5   Explain briefly the effect that each of the following material defects can have on
    finished work

    a) knotholes on surface of shuttering plywood   (1 mark)

    b) sawn edges on architrave   (1 mark)

    c) twisted lining jambs   (1 mark)

    d) extra tight tongued and grooved flooring boards.   (1 mark)

6 With the aid of a sketch, show a method of repairing a rotten part at the end of a door jamb. (4 marks)

7 State why, with regards to a fixed circular saw

a) a push stick should be the right length (2 marks)

b) non-slip floor surfaces must be in front and behind the saw bench. (2 marks)

8 Show, by means of a sketch, how the pipe shown in the figure below can be insulated against sound and boxed in to provide flush surfaces.

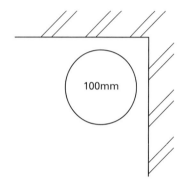

100mm

9 State **two** precautions that should be taken when handling and erecting truss rafters. (4 marks)

10 Explain, with the aid of sketches, how

a) the accuracy of a spirit level can be checked (2 marks)

b) to transfer levels over a 6m distance with a 1m level. (2 marks)

11 State the name of the following components:

a) the horizontal member at the bottom of a doorframe (1 mark)

b) the horizontal board to which gutters are fixed (1 mark)

c) vertical members fitted between treads on a staircase (1 mark)

d) shaped blocks at the base of vertical architraves. (1 mark)

12 Explain

a) why a glued end-to-end joint would fail when subjected to tension (2 marks)

b) with the aid of sketches, how the joint strength could be improved. (2 marks)

13 A carpenter and his trainee took nine hours laying laminate flooring. Using the following data, calculate the **total** cost to the customer

    a) carpenter's rate   = £14.50 per hour

    b) trainee's rate   = £6.35 per hour

    c) materials   = £387.67

    d) overheads and profit   = £267.29.   (4 marks)

14 Sketch sections through each of the following timber profiles:

    a) taurus   (1 mark)

    b) ogee   (1 mark)

    c) scotia   (1 mark)

    d) shiplap.   (1 mark)

15 If timber is sawn to the wrong sizes and delivered to site, explain how this can affect a job's profit margins and deadlines.   (4 marks)

16 Explain briefly why fixed circular saws have

    a) gullets on the saw blade   (1 mark)

    b) crown guards   (1 mark)

    c) take-off extension tables   (1 mark)

    d) set-on teeth.   (1 mark)

17 Sketch a section through a hollow suspended timber ground floor and dwarf sleeper wall, naming all the parts.   (4 marks)

18 State **two** reasons why boring and notching joists for other trades must not exceed the requirements of the current Building Regulations.   (4 marks)

19 Explain why the bottoms of floor-mounted kitchen units normally stand adequately clear of the floor on adjustable legs.   (4 marks)

20 Sketch a section through a storm-proof window sash showing the window head, hinge, sash head and glazing.   (4 marks)

abbreviations	shortened versions of words or names, often using just the initial letters of each word
acoustic	to do with the way sound travels around
amenities	facilities such as toilets, rest areas, etc.
asbestosis	a serious lung condition caused by breathing in asbestos
ballistic tool	any of a range of tools where the mechanism involves a 'throwing' action
balustrade	unit comprising handrail, newels and the infill between it and the string, which provides a barrier for the open side of the stair
barrier cream	a protective cream that stops water and other substances getting through
beam compass	a special carpenter's compass for creating arcs and circles with a wide diameter
blemish	a stain, scar or spot
'bottleneck' effect	when things get jammed, especially when a large flow comes into a narrow passage and not all of it can get through, as in the neck of a bottle
clout nail	a large-headed nail
combustion	burning, the action of fire
compliance	obeying, fitting in with
conservation	preservation of the environment and wildlife, or of rare or special buildings
contamination	when harmful chemicals or substances pollute something (e.g. water)
contingencies	plans set up just in case something happens
corrosive	a substance that can damage things it comes into contact with (e.g. material, skin)

counterbalance	a weight used to 'counter' or act against the weight of something else, so that it doesn't fall over
cut-outs	scaled-down shapes that represent bigger things, which can be used on a scale drawing to help you plan
decibel (dB)	the standard unit for measuring noise level
dermatitis	a skin condition where the affected area is red, itchy and sore
discrepancies	when there is a difference or variation between two things that should be the same
dismantle	take apart, take down carefully
DPC	damp proof course, a substance that is used to prevent damp from penetrating a building
disproportionate	out of proportion to something else, far more or less than you would expect
duration	how long something goes on
electrocution	death through coming into contact with an electric current
employer	the person or company someone works for
employee	the person employed by the employer, the member of staff
exhaustive	absolutely complete
going	the depth of a step (the measurement from a step's riser to the edge of the step)
halving joint	the same amount is removed from each piece of timber so that when fixed together the joint is the same thickness as the uncut timber
Health and Safety Executive (HSE)	government organisation that enforces health and safety law in the UK
hydraulic	worked by water or other liquid in pipes
hypotenuse	the side of a right-angled triangle opposite the right angle
impregnated	soaked right through
inconsistencies	when things are not the same, not consistent

in situ	on the spot, in the actual location where it is needed
institutional	to do with an institution such as a hospital or school
interim	in the time between, for the time being, as a holding measure
intersection	the point at which things join or cut across each other
jeopardise	endanger, put at risk
legislation	laws or the making of laws
manifestation	how something looks or is presented
multiples	sets of more than one; in housing, sets of matching dwellings that are all the same (first occurrence)
muntin	
node	a key point (especially on a critical path chart)
nogging	a short length of timber, most often found fixed in a timber frame as a brace
no-show	when someone does not turn up as planned
objectives	aims, purposes
out of square	not properly at right-angles, a bit skew
paramount	the most important thing, the greatest concern, above all others
penalty clause	a clause in a contract saying a fine has to be paid, or some other penalty made, if a certain thing happens, e.g. the job overruns
prevalent	common, often found
prohibition	a ban, saying something cannot happen or be done
projection	sticking out, a part that juts out
prosecute	take someone to court for committing a crime
prospective	likely or possible in the future, but not actually happening or approved now

radius rod	a rod which restricts the movement of a part to a given arc
rectify	put right
recurrence	happening again
remit	scope, job, the areas an organisation or individual has to cover
residential	where people live, rather than a business district, for example
reverberation	echoing, reflecting sound and vibration
roofing ready reckoner	a set of mathematical tables giving a quick way to work out rafter lengths
sash	part of an old-fashioned style of window with a top and a bottom section that can be moved up and down with cords
segmental	made or divided into sections, usually equal sections
serpentine	like a snake, curving one way and the other
serrated	with teeth on, like a saw
silicone	a chemical used in rubbers, seals and polishes that is waterproof, very stable and long lasting
skew-nailed	nailed with the nails at an angle
solvent	a substance that dissolves another e.g. paint stripper
stipulation	a condition of an agreement, a particular term of a contract
string	main board to which treads and risers are fixed
superseded	overtaken by, replaced by
surveillance	carefully watching over or keeping an eye on
susceptible	prone to, vulnerable to
sustainability	the ability to last or carry on, how easy something is to keep going
tapered	getting thinner towards one end
telescopic	sliding or arranged like the joints of a telescope, so it can be packed away small

toxic	poisonous	**ventilation**	to do with air going in and out
tusk tenon joint	a kind of mortice and tenon joint that uses a wedge shaped key to hold the joint together	**vibration white finger**	a condition that can be caused by using vibrating machinery (usually for very long periods of time). The blood supply to the fingers is reduced which causes pain, tingling and sometimes spasms (shaking)
UPVC	unplasticised polyvinyl chloride, a relatively stiff material, sometimes brittle in cold weather, of which sewer pipes or water pipes are made.		
		vice versa	the other way round
vacuum	a space that is completely empty, with all the air sucked out	**WBP**	weather/water boil proof